Antiques at a Glance
CERAMICS

Antiques at a Glance

CERAMICS

James Mackay

This edition published 2002 by
PRC Publishing Ltd,
64 Brewery Road, London N7 9NT

A member of Chrysalis Books plc

Produced for Greenwich Editions
64 Brewery Road, London, N7 9NT

A member of Chrysalis Books plc

© 2002 PRC Publishing Ltd.

All rights reserved. No part of this publication may be reproduced, stored in a retrieval system, or transmitted in any form or by any means, electronic, mechanical, photocopying, recording, or otherwise, without the prior written permission of the Publisher and copyright holders.

ISBN 0-86288-453-5

Printed and bound in China

All the images were kindly supplied by © Christie's Images Ltd 2002.

Contents

Introduction 6
Delftware, Maiolica and Faience
    Deruta Dish 26
    Istoriato Dish. 27
    Pesaro Maiolica Istoriato Dish. 28
    Urbino and Deruta Wares 29
    Famiglia Gotica Albarello 30
    Castelli Albarello 31
    Spirit Flask 32
    Castelli Plaques 33
    Pesaro Maiolica Dish 34
    German Faience and Dutch Delft Group 35
    Dutch Delft Jars and Bottles 36
    Dutch Delft Vases 37
    English Delftware Group 38
    English Delftware Group 39
    English Delftware Group 40
    Henry Delamain Wares 41
    Dublin Delftware Plates 42
    Nuremberg Plaque 43
    French Faience Tureen 44

Oriental Porcelain
    Kakiemon Dragon 45
    Imari Vase. 46
    Imari Dish 47
    Cloisonné Vase 48
    Cloisonné Vase 49
    Famille Rose Vase 50
    Famille Rose Vase 51
    Famille Rose Bowl 52
    Famille Rose Hollow Wares 53
    Famille Rose Wares 54
    Famille Verte Elephant and Figures 55
    Yuan Jar 56
    Dragon Jar 57
    Yaoling Zun 58
    Famille Verte Stembowl 59
    Famille Verte "Birthday" Dish 60

- *Famille Verte* Wares ... 61
- *Famille Verte* Group ... 62
- *Famille Verte* Vases ... 63
- Ming-style "Lotus Bouquet" Dish ... 64
- Ming-style Moonflask ... 65
- Ming-style Lotus Dish and Stemcup ... 66
- Chinese Porcelain ... 67
- Chinese Porcelain Vases ... 68
- Baluster Jars and Covers ... 69

European China
- Meissen Foliate Dish ... 70
- Meissen Porcelain Trifles ... 71
- Meissen Porcelain Trifles ... 72
- Meissen Snuffbox ... 73
- Meissen Snuffboxes ... 74
- Meissen Armorial Dish ... 75
- Böttger Hausmalerei Beaker ... 76
- Möllendorff Dinner Service ... 77
- Meissen Dinner, Dessert and Side Plates ... 78
- Four Meissen Figurines ... 79
- Meissen Part Dinner Service ... 80
- Meissen Dishes ... 81
- Meissen Group ... 82
- Hofservice Dish ... 83
- Scent Bottle ... 84
- Naples Coffee Cup and Saucer ... 85
- Doccia Snuffbox ... 86
- Du Paquier Tableware ... 87
- Vincennes Miniature Cabaret ... 88
- Sèvres Armorial Dessert-plate ... 89
- Sèvres Tablewares ... 90
- Sèvres Plates ... 91
- Sèvres Serving Dish ... 92
- Dihl and Guerhard Plate ... 93
- Gardner and Popov Porcelain Figures ... 94
- Gardner Porcelain Figures ... 95
- Mettlach Plate ... 96
- Villeroy & Boch Vases ... 97
- Clement Massier Jardinière ... 98

British Pottery and Porcelain
- Staffordshire Slipware Jug ... 99
- Staffordshire Hawks ... 100
- Staffordshire Cow Creamers ... 101
- Pratt Lion ... 102
- Sir Isaac Newton Bust ... 103
- Staffordshire Pearlware Group ... 104
- Staffordshire Watch Holder Group ... 105
- Chelsea Jug ... 106
- Chelsea Bocage Imperial Shepherds ... 107
- Bow Group ... 108
- Chelsea Coffee Pot ... 109
- Derby Figures ... 110
- Derby Teacup, Saucer and Sugar Bowl ... 111
- Lowestoft Cream Jug, Coffee & Teapots ... 112
- Worcester Deep Plate ... 113
- Worcester Serving Dish ... 114
- Worcester Tableware ... 115
- Worcester Dessert Service ... 116
- Pearlware Models ... 117
- Bristol Plaque ... 118
- Coalport Dessert Service ... 119
- Minton Teapot ... 120

Art and Studio Pottery
- Linthorpe Vase ... 121
- Ault Vase ... 122
- Doulton Lambeth Bowl ... 123
- Doulton Lambeth Vase ... 124
- Burmantofts Toad Spill Vase ... 125
- Belleek Elephant ... 126
- Martin Brothers Jug, Vase and Birds ... 127
- De Morgan Vase ... 128
- Burmantofts Faience Vases ... 129
- Moorcroft Vases ... 130
- Poole Pottery Dish and Spill-holder ... 131
- Rookwood Vases ... 132
- Roseville Twin-handled Bowls ... 133
- Weller Dickensware Vase ... 134
- Grueby Pottery Vase ... 135
- Amphora Bowls, Vases and Jugs ... 136
- Bernard Leach Box and St Ives Chargers ... 137
- Shoji Hamado Dish ... 138
- Bernard Leach Charger ... 139
- St Ives Plate ... 140
- Hans Coper Dish and Vase ... 141
- Hans Coper Vase ... 142
- Clarice Cliff Dinner Service ... 143

Index ... 144

Introduction
The History of Ceramics

Humanity is perhaps not unique in the habit of storing up objects for their own sake, but is the only animal capable of translating material possessions into other values and assessing their worth in terms of hard cash.

According to the *Encyclopaedia Britannica* "antique" means "old" but also carries connotations of esthetic, historic or financial value. Formerly the term was applied to the remains of the classical cultures of Greece and Rome, now more specifically labeled as "antiquities," but gradually "decorative arts, courtly, bourgeois and peasant, of all past eras" came to be considered antique.

This definition is somewhat vague, though it hints that mere age alone is not sufficient to make an object worthy of the appellation "antique." *The Oxford English Dictionary* is even more vague in its definition: "Having existed since olden times; of a good age, venerable, old-fashioned, antiquated such as is no longer extant; out of date, behind the times, stale; of, belonging to, or after the manner of any ancient time; a relic of ancient art, of bygone days."

The legal definition of an antique also varied considerably from one country to another. The United Kingdom Customs and Excise Tariff Act of 1930 specified that objects manufactured before 1830 (i.e. a hundred years old or more) would be regarded as antique and therefore exempt from payment of duty on import. The United States Tariff Act of 1930 exempted from duty "Artistic antiquities, collections in illustration of the progress of the arts, objects of art of educational value or ornamental character... which shall have been produced prior to the year 1830." Across the border in Canada, however, their Customs Tariff Act of 1948 defined as antiquities "all objects of educational value and museum interest, if produced prior to 1st January 1847."

Legal definitions, originally designed to cover material a century old, became fixed in such a way as to exclude anything produced after the Regency period. It was often stated that the main reason for adhering to the year 1830 was the fact that craftsmanship deteriorated after that date. For this reason bodies, such as the British Antique Dealers' Association, clung to 1830 as the chronological criterion in defining an antique.

As time passed, however, this inflexible ruling seemed more and more untenable from a purely legal viewpoint, and for the purpose of the avoidance of the payment of import duty the date criterion was subsequently modified in certain countries. The United Kingdom Customs and Excise tariff act of 1959, for example, outlined that duty would not be payable on the import of objects, "if manufactured or produced as a whole, and in the form as imported, more than a hundred years before the date of import." More

RIGHT: Glazed earthenware Persian vase.

INTRODUCTION

recently, the United States customs adopted a straightforward hundred-year rule, and with the advent of Value Added Tax and similar imposts in other countries, the hundred-year rule is now widely observed.

In periods of economic uncertainty interest in collectable objects tends to increase at the expense of more traditional forms of investment. What used to be the principal motive in collecting—a need to identify with the past—has tended to yield to a need for the form of security that the possession of tangible objects brings when money itself is diminishing in worth and true meaning. It becomes less important to the collector to amass objects of great antiquity, especially since the supply of antiques has become scarcer due to worldwide demand, and leads the collector to turn inevitably to more recent products. For those reasons, many dealers in the 1970s and 1980s adopted a fifty-year rule, so that objects of the 1920s and 1930s could be encompassed. First the artifacts of the late-Victorian and Art Nouveau periods became perfectly acceptable, and were then rapidly followed by Art Deco and the products of the immediate prewar period. Now date lines have been virtually abandoned as collectors and dealers discover newer and newer objects on which to focus their attention.

Greater flexibility in defining what might be regarded as an antique came not a moment too soon, as the supply of fine-quality pieces produced before 1830 had all but dried up by the 1960s. At the same time, many museums with seemingly unlimited funds were intent on expanding their collections. This meant that the amount of quality antiques available to the market began to dry up, as well as having the unfortunate effect of pushing up the market value of what material was left. Some categories of antiques, such as 16th- and 17th-century silver, early Meissen and Chelsea porcelain, and Ravenscroft glass, soon went beyond the reach of all but the wealthiest collectors. Not only did museums tend to create a shortage of material available to the private collector but, by imaginative and intelligent use of their acquisitions, they heightened the interest of established collectors and laymen alike, thereby increasing the demand for antiques still further.

While this tendency was gathering momentum in the 1960s and 1970s, the growing interest in the private sector was inevitable anyway. A higher level of general prosperity and higher standards of education were only two of the factors that made the public not only more appreciative of all that was best from times past, but also gave them the money to indulge their tastes. Traditionally antique collecting had been the closed preserve of the upper classes, who had the money, the education, and the social background to indulge their tastes. Collecting antiquities developed in Renaissance Italy and spread slowly to more northerly countries. In Britain, for example, the fashion for collecting things of the past only really began to develop in the late 18th century. Emphasis was laid on classical antiquities, fostered by the classical education of the times as well as the acquisitive habits picked up during the obligatory Grand Tour.

From then on, an antiquarian interest in the material objects of the past tended to lag behind by one or two centuries. Collectors of the Regency era discovered an interest in Tudor and Jacobean furniture, the Victorians looked to the Restoration and early Georgian period, and the Edwardians had the highest regard for the products of the 18th and early 19th centuries. In general, collectors and cognoscenti alike disregarded the products of their immediate forebears, which explains why the 1830 rule endured as long as it did.

Interest in collectables has also gone in cycles. There are five- or ten-year highs and lows in different categories, as well as periods of slump due to socio-political and economic factors. Antiques that came on to the market in 1917 or 1942, for example, could be picked up for a song. However, people didn't have the money to spend in those years, and there was a general reluctance to invest in material that might be destroyed by enemy action or plundered in the uncertainties of war.

Conversely, the economic upheavals of the late 1960s and early 1970s led to a flight of money out of traditional forms of investment, such as stocks, shares, unit trusts, real estate and building societies, and into art and antiques. The devaluation of sterling in November 1967 and the subsequent run on the dollar and then the franc created a wave of near hysteria in the antique markets of Europe and America. The leading salerooms reported a 50 percent increase in turnover in the ensuing 12 months alone, right across the board, although in certain categories the turnover was up by as much as 100 percent (for prints and drawings), while silver sales increased by 69 percent.

Coupled with this astonishing increase in sales, it was significant that the leading salerooms on both sides of the Atlantic began diversifying into material of more recent vintage than was generally accepted as antique, and this trend has steadily developed ever since. This even led to the development of separate auction houses, such as Christie's in South Kensington, London, catering specifically to the "newer" antiques, including many articles that would not have been regarded as collectable a few years previously.

The major salerooms, as well as the multitudes of lesser auction houses, were only encouraging a trend that was already there. In the same way, the junk shop of yesteryear has been elevated to the antique shop of today; the weekend junk stall or barrow in a street market has become a booth in a permanent antique market; and the better pieces traded in car-boot sales rapidly move up the scale, with a corresponding mark-up in price at each stage.

As the amount of quality antiques available to the market dwindled, it seemed paradoxical that antique shops were proliferating everywhere at an astonishing rate. The number of good antique shops remained fairly static, nor did they find it any easier to obtain quality material for their stock. The answer to this paradox was, at first, a general lowering of standards; if you clung to increasingly untenable date lines, whether 1830 or 1870, then it was inevitable that you had to settle for second-best or some sacrifice in workmanship, condition or quality.

The more astute collectors ignored date lines and explored the potential of later material. If the products of the Baroque, Rococo and Neo-classical periods were no longer available, might not there be much to commend in, say, the products of the Second Empire in France, the Biedermeier era in Germany or the Victorian era in general? It was fashionable at one time to write off the entire Victorian period as one of uniformly bad taste. Much of the opprobrium heaped on the bourgeois fashions of the 19th century by subsequent generations was undeserved. It is true that in furniture and art, as in the material comforts of everyday life, the Victorians showed a predilection for the massive, the ornate and the fussy; but not all was tawdry or tasteless by any means.

That the Victorians were capable of perpetrating, and apparently enjoying, objects of unbelievable hideousness is true, but at the same time there were serious attempts to raise standards. The much-abused

ANTIQUES AT A GLANCE: CERAMICS

10

Great Exhibition of 1851 did more than is often realized to encourage pride in craftsmanship and demonstrate that a thing could be beautiful as well as functional. Though much that is Victorian was, not so long ago, regarded as hardly worth preserving, there were many other things that possessed enduring qualities, and were recognized as such by discerning collectors, long before such objects had earned the title or dignity of antique. It has to be added that even the fussy and the florid, the over-ornamented and the downright ugly from that much-maligned period have acquired a certain period charm. Truly, distance does lend enchantment.

Conscious efforts to improve public taste and foster pride in workmanship seem like oases in the wilderness of materialism and mass production. In England, the Arts and Crafts Movement inspired by William Morris in the 1880s was an attempt to recapture something of the primeval simplicity in craftsmanship—a reaction against the pomposity and over-ornateness of Victorian taste. It was a precursor of that curious phenomenon at the turn of the century known as Art Nouveau in Britain, as *Jugendstil* ("youth style") in Germany, or as Liberty style in Italy (from the well-known London department store which was one of its great proponents).

The practitioners of the New Art went back to nature for inspiration, and invested their furniture, glass, silver and ceramics with sinuous lines and an ethereal quality. In turn, this provoked a reaction that resulted in the straight lines of the Bauhaus and the geometric forms associated with Art Deco in the inter-war period. It has to be admitted that these styles and fashions seemed ludicrous to many people at the time, especially in their more exaggerated forms; nevertheless they were the outward expression of a minority in art, in architecture, in furniture, textile and ceramic design, which strove for improvements (as they saw it) in the production and appearance of objects. These were not only objects intended purely for decorative purposes, but those used in every phase and aspect of life.

The products of the Arts and Crafts Movement, of Art Nouveau and Art Deco were despised and neglected in succession, and then, after a decent lapse of time, people began to see them in their proper perspective and appreciate that they had a great deal to offer to the collector.

Nevertheless, it is also fair to comment that the century after 1830 was a barren one as far as the production of fine-quality material was concerned. Thirty years ago collectors made a fine distinction between what was merely old but had no particular merit on grounds of esthetic features or workmanship, and those objects which had some qualities to commend them. Nowadays, however, as demand continues to outstrip supply, there is a tendency to talk up wares that may be old, but are commonplace and mediocre nonetheless. The insatiable demand especially at the lowest end of the market decrees that this should be so, but it is important for the collector to discriminate and learn to recognize the features and factors that distinguish the worthwhile from the second-rate. Ultimately these are the factors that govern the soundness of any investment in antiques.

In the course of this century there have been startling developments in education, communications, travel and living standards. People are generally more

LEFT: Meissen figures of Pantalone and Scapin from the Commedia dell arte series.

affluent today than were their parents or grandparents. They enjoy shorter working hours and a larger surplus disposable income. Through education and such external stimuli as the cinema and television programs about antiques and collecting them, a lot more people have a greater awareness of things of beauty or of antiquarian interest. More and more people now have the time and the money to indulge in collecting objects, which in the past was the preserve of a privileged few.

A greater general awareness of what is beautiful and worthwhile inevitably tends to encourage better craftsmanship. Despite the general perception of the period between the two world wars as the nadir in fine design and workmanship, there were also many individuals and groups who were active in Europe and America in promoting design consciousness. Today, the products of their studios and workshops, especially in the fields of furniture, ceramics, glass and metalwork, are deservedly sought after and fetch correspondingly high prices. This trend has continued to the present day, with the result that each year the artifacts created by the most imaginative and innovative graduates of the art schools and colleges are eagerly snapped up as the antiques of the future.

Britain, which led the way in the mid-19th century, also pioneered attempts to foster good design in an infinite range of useful articles, from household appliances to postage stamps, through the medium of the government-sponsored Design Centre and the Council of Industrial Design. During World War II, when there was a shortage of materials and manpower, these schemes helped to develop the utility concept, which extended over the entire range of manufactured goods. At the time, "utility" was often equated with shoddy and second-rate, but in more recent times collectors have begun to appreciate the simple lines of the applied arts of the so-called austerity period.

There was a time when objects were collected for their own sakes; as examples of exquisite craftsmanship, beauty or rarity. Perhaps the reason for collecting was nothing more than the charm of owning something of great age. At any rate, intrinsic worth was seldom of primary consideration. Nowadays, however, there is a tendency for the collector to be aware of values and to prize his possessions not only for their esthetic qualities, but also as investments.

Gone are the days of the great gentleman-collectors, such as Sloane, Cotton, Harley, Hunter, Hearst and Burrell, whose interests covered every collectable medium and whose tastes were equally developed for paintings and incunabula as for coins and illuminated manuscripts.

Even the computer billionaires of the present day could scarcely emulate the feat of the late Andrew Mellon, who in the 1920s once purchased 33 paintings from the Hermitage for $19 million. But while there are very few private individuals, who could now afford to buy a Leonardo, a Rembrandt or even a Van Gogh, there are millions of people throughout the world who have the leisure to specialize in some chosen field, and the surplus cash to acquire the material for their collections. There are countless aficionados who have formed outstanding collections of porcelain, silver, prints or glass, who have specialized in the products of individual potteries, or Depression glass, or Kilner paperweights or Goss china.

At the lower end of the spectrum there are hundreds of different classes of collectable, from the frankly ludicrous, such as bricks, barbed wire and lavatory chain-pulls, to the fetishistic, including whips and certain articles of ladies' apparel. There are also collectors of commemorative wares and even objects associated with one's profession, such as dental and medical instruments. The collecting virus is now endemic and insatiable.

Pottery and Porcelain

For all practical purposes, the earliest kind of pottery, which is still available to the collector, is maiolica. This earthenware (which should not be confused with the rather similarly spelled majolica of the 19th century), usually found in the form of drug jars, dishes, jugs and flagons, with polychrome decoration, was produced in Urbino, Faenza and other Italian towns and exported to northern and western Europe. Good quality maiolica is scarce and worth a fortune when it passes through the saleroom, although from time to time items of lesser quality can still be picked up at fairly reasonable sums.

In ceramics, Italian maiolica vied with early French faience, but by the end of the 16th century Delft in Holland was an important center for tin-glazed earthenware. This was mainly blue and white in decoration in imitation of the Oriental porcelains, which were now beginning to be imported through The East India Company and its European rivals. In the early 17th century, delftware became a generic term for decorative earthenware manufactured at London, Liverpool, Bristol and Wincanton in England as well as at Dublin, Limerick and the aptly named Delftfield in Glasgow, each area producing its own distinctive variation. The blue-dash chargers of the late 17th and early 18th centuries are much sought after, not only on esthetic grounds but also for their historic significance, as they are often decorated with portraits of royalty. Mugs, beakers, barbers' basins, bleeding bowls and apothecaries' jars as well as rack-plates provide ample scope for the collector. Much of this material belongs to a later period, delftware remaining fashionable until the late 18th century.

Styles in Britain as the 17th century drew to a close were strongly influenced by the Baroque fashions, which originated in Italy and spread to Spain, France and Germany. These were largely introduced to England by the influx of Huguenot refugees after the revocation of the Edict of Nantes in 1685 increased the persecution of non-Catholics. Baroque (literally the French word for "irregular") was characterized by asymmetrical, curving lines but gradually developed into extravagant and elaborate ornament, festooned with ribbons, scrolls, swags, shells, cornucopias and cupids.

During the same period English pottery began to compete successfully with imported wares, although there was nothing to equal the hard-paste porcelain that flooded in from China. The blue and white motifs on Chinese ceramics triggered off a craze for *chinoiserie,* which influenced other forms of the applied arts, from furniture to textiles and metalwork.

Britain emerged as a world power in the early 18th century, after the brilliant successes of its forces in the War of the Spanish Succession. In later conflicts Britain rivaled France for the mastery of the colonial world, in India and the Western Hemisphere and emerged triumphant in 1763. The accession of the Hanoverian kings in 1714 increasingly involved Britain in European politics and artistic influences. Despite the Jacobite rebellions of 1715 and 1745, Britain enjoyed a long period of relative stability and rising prosperity. Greater affluence was reflected in the furniture, silverware and decorative arts of the period.

The Georgian era is conveniently divided into Early Georgian, covering the reigns of the first two Georges (1714-60), and Late Georgian, corresponding with the long reign of George III (1760-1820). The era as a whole witnessed a tremendous development in architecture, which in turn influenced styles in the applied and decorative arts. The Early Georgian period coincided with the zenith of the Baroque in Europe, with its emphasis on curves and scrolls in everything from the legs of tables to the handles of coffeepots. Scallops and acanthus leaves decorated the corners and joints of furniture as well as the rims of vessels.

The craze for curved lines culminated in the 1730s with the rise of Rococo, a much lighter, more delicate style than Baroque and clearly a reaction against its tendency to the massive and fussy. The Italianate word was actually derived from two French terms—*rocaille* (rockwork) and *coquille* (shell). It arose out of the vogue for grottoes in landscape gardening and was characterized by floral swags and garlands as well as "C" and "S" curves in great profusion. Britain lagged behind the Continent so it was not until the middle of the century that the Rococo fashion reached its height in England. It lent itself very well to setting off Oriental motifs, later joined by Indian art forms and continued, in a more restrained form, right through to the end of the 18th century. In general, Rococo represented a much lighter approach to form and decoration than the Baroque. By the end of the century, however, styles were becoming more eclectic, often blending the Rococo with the Neo-classical and even the Gothic, an artificial revival of certain medieval forms, such as pointed arches.

Meissen and Sèvres pioneered European porcelain in the first half of the century but it was not until 1750 that the manufacture of similar wares began at Derby, followed by Worcester (from 1751), Chelsea (1744–70) and Bow (1744–76). English bone china rivaled the French and German products in its uniformly high quality in composition, decoration and potting. There was, of course, a penchant for Chinese and Japanese decoration but European flowers, penciled motifs, armorial features and transfer-printed ornament were also popular. The shapes favored by the potters were closely modeled on the rococo styles used by contemporary silversmiths, but there was an enormous demand for figurines, groups and centerpieces.

In ceramics English craftsmen were making enormous strides and creating distinctive innovations. This was the era of Josiah Wedgwood, pioneer of Queen's Ware for tea sets and other tablewares. Later he turned his attention to various forms of stoneware, first black basaltes and later jasperware in which white cameos and reliefs were set against a ground of pale pastel colors. Both materials were utilized in the production of a huge range of useful and ornamental pottery, which laid the foundations of timeless classical lines that remain popular to this day. Many potteries sprang up in this period, competing with each other in the fields of earthenware and soft-paste porcelain. Today, the products of Coalport and Caughley, Lowestoft and Longton Hall, Leeds, New Hall, Rockingham, Spode, Swansea and Minton all have their devotees.

Regency fashions did not die out abruptly with the accession of William IV in 1830, but already great social and political changes were sweeping over Britain. Although they were not crystallized until the middle of the century, it is customary in the world of fashion to

RIGHT: A Deruta dish c.1500 painted in blue, yellow, orange, and copper-green. The central raised boss, ringed in blue, has a portrait of a young woman in Renaissance dress and cap.

INTRODUCTION

speak of the Victorian era as if it had commenced seven years before the young queen ascended the throne in 1837. While fashions in ceramics, glass and silver did not change much before the Great Exhibition of 1851, the styles, techniques and even the materials of furniture had been undergoing radical alteration in previous decades. The rapid growth of economic prosperity, which came in the 1830s, stimulated a tremendous demand for furniture.

Although Victorian is used as a generic term for the applied and decorative arts of the 19th century, it had its European counterparts. The Biedermeier furniture and decor of mid-19th century Germany was long derided as conventional and unimaginative. Its name was derived from a fictional character, Gottlieb Biedermeier, a rather simple-minded, essentially philistine petit-bourgeois; an image that added to the derision surrounding it. But today its very solidity is now regarded as highly commendable, while there is much to delight the eye in many of the lesser pieces, especially the ceramics, glass and silver. The French equivalent was Second Empire, roughly contemporary with the reign of Napoleon III (1852–71), and likewise unfairly dismissed by the generations that followed immediately.

As the 19th century drew to a close, influences and developments in the applied and decorative arts became much more cosmopolitan. The French expression *fin de siecle* has come to be synonymous with decadence. The estheticism expressed by Oscar Wilde, J.K. Huysmans and Robert de Montesquieu had its parallels all over Europe and even extended to America. The strange, exotic, luxuriant and faintly decadent spirit of the times had its flowering in the sinuous lines of Art Nouveau. It was a period of eclecticism, when artists and designers drew freely on all the artistic styles and movements of previous generations from every part of the world, and often jumbled them together in a riotous compote. The craze for Japanese art and artifacts was predominant, but inspiration was also derived from the ancient civilizations of Greece and Rome, Persia, India, Peru, Mexico, Benin and China. Nevertheless, it is significant that the German term for the turn of the century developments in the arts was *Jugendstil* or "youth style," implying vigor, freshness, originality and modernity. The followers of the Arts and Crafts Movement, on the one hand, rejected modern mass-production techniques and sought to return to first principles, to handicrafts and inspiration from nature. The disciples of *Jugendstil*, on the other hand, did not spurn the machine if it could be used to their advantage, and they looked forward, in an age of speed and light, to producing works which would express the qualities of the age.

The major countries of Western Europe and the United States each had an important part to play in the development of the applied and decorative arts. In Britain, the desire for improvement in industrial design can be traced back to the Great Exhibition of 1851 and, even earlier, to the Royal Commission on the Fine Arts in 1835. The Gothic Revival of the mid-19th century stimulated interest in medievalism, reflected in the religious overtones of the early work of William Morris, Philip Webb and Edward Burne-Jones.

It would be difficult to overestimate the importance of Morris to the artistic development of Britain in the late-19th century. Two important movements stemmed directly or indirectly from his teachings. One was the Arts and Crafts Movement of the 1880s, which aimed at bringing artists and craftsmen closer together, to raise standards of workmanship, and to put artistic pride into even the most mundane articles. The other was the Aesthetic Movement, founded on an elitist principle, which genuinely strove to raise standards of design and taste. On the continent of Europe, the styles that culminated in Art Nouveau had their origins in France where two major artistic movements flourished in the last third of the century, Naturalism and Symbolism. There were parallel developments in Belgium, Italy and Spain, which fused in the 1890s, and were enthusiastically adopted in England and given a distinctly British flavor before finding their way back across the Channel in the guise of *le style anglais*.

The Civil War of 1861–65 was no less traumatic for the United States than the German occupation and the Commune of 1870–71 were for France. The rapid expansion of industry, coupled with widespread immigration from Europe, changed the character of the country in the last three decades of the century. America ceased to be a pioneer land and in the aftermath of the Spanish-American War of 1898 assumed an imperial role. In the arts, as in politics, America now reached out to every part of the globe. Interest in the arts of China and Japan, of Latin America and Africa, were combined with the traditional styles, which were themselves derived from the British, Dutch and German of the colonial era, or imported with the waves of European migration from the 1860s onwards. This blend of Oriental or pre-Columbian influences with the styles and techniques of Europe could be seen in the furniture, glassware and ceramics of America at the turn of the century. These, especially art glass and studio pottery, found their way to Europe where they exerted a considerable influence on the applied arts of the present century.

The young architects and designers of the Chicago School revolutionized the design of buildings and furniture from the 1890s onwards. In Europe, the break with the old ideas was often more dramatic, as, for example, in the Sezession movement in Austria and Germany. The world of the arts and crafts was thrown into turmoil. Many new ideas and styles appeared; some were short-lived and have now become crystallized in the history of the period, but others contained the seeds that germinated in the 1920s and came to full maturity nearer the present day.

Notably, the Bauhaus movement in Germany influenced the development of Art Deco in the 1920s, with its rejection of the curvilinear extravagance of Art Nouveau. Geometric forms and bright primary colors were in tune with the Jazz Age. Many new forms emerged in the inter-war period; the cigarette lighter and the powder compact, for example, replaced the snuffbox and vinaigrette of earlier generations as ornamental objects which have since become eminently collectable.

Pitfalls and Plus Factors

Collectors, and not always beginners by any means, are often puzzled by the vast price differential between two objects, which are superficially similar. In some cases there may be as many as a dozen criteria governing the value of an object. These could be age, materials, type of construction, quality of craftsmanship, artistic or esthetic considerations, unusual technical or decorative features, the provenance of personal association, the presence or absence of makers' marks, full hallmarks, dates, and inscriptions. These and other criteria vary in importance from one object to another, and may even vary within the range of a single category, at different periods or in certain circumstances. Visiting museums and stately homes or handling actual objects at sale previews, as well as studying all the available specialist literature on any given subject, will help the aspiring collector

to get a "feel" for the subject, but there is no shortcut to gaining expertise.

Above all, condition is the most problematic factor in assessing the worth of an object. Reasonable condition, of course, depends on the object and the degree to which damage and repairs are accepted by specialist collectors and dealers. A very fine early paneled chair might well have a replacement to the last two inches of a back leg and if well done this would have practically no effect on the price. On the other hand, a run-of-the-mill lacquer object, scratched and crudely repainted, would be almost valueless. The general rule is that where a piece is interesting and few collectors have one in their collection, a much damaged example will fetch a surprisingly good price. This often happens with early examples of ceramics from important factories, whereas a common jug, missing a handle, will be virtually worthless. The failure to appreciate the effects on value of poor condition, loss of original surface, fading or rubbing, is one of the most common causes of the misunderstanding that arise between collectors and dealers.

The market value will also take into account the imponderables of where, when and how an article came on to the market. There is often a wide disparity in the sum that identical objects may fetch in a London saleroom, in a provincial auction or a country house sale. The individual vagaries of obsessiveness of two or more wealthy private collectors may grossly affect the auction prices of certain objects on a particular occasion, while absolutely identical objects can (and sometimes do) fetch half these sums at other times in other places. Moreover, there is both a greater disparity between prevailing auction realizations and dealers' retail prices in general, and between the prices of one dealer and another, who are not always situated miles apart! Unfortunately, the collector cannot shop around before making a purchase. There are still bargains to be picked up; but all too often one finds that objects are outrageously over-priced in general antique or junk shops. Contrary to popular belief, some of the keenest bargains are still to be found in metropolitan antique shops and markets, where competition comes into play; conversely some of the most atrocious over-pricing has been observed in provincial towns or the antique "boutiques" in tourist areas. There is no clearly definable regional pattern of pricing in the United Kingdom or the United States or anywhere else for that matter; this is something that collectors have to explore for themselves.

While a certain amount of judicious repair and restoration is permissible, fakery is reprehensible and usually detracts from whatever value the genuine part of the object may have had before it was tampered with. Unfortunately the dividing line between legitimate repair and outright faking is often a rather tenuous one; but the general principle is that any deliberate altering of an object to create something of greater value is a form of fraud. It occurs most often in furniture where large but unfashionable and unsaleable pieces are dismembered and their timbers used to re-create small pieces which, with a bit of luck, can be passed off as genuine articles. The other problem, which besets the unsuspecting collector, is reproduction. Although early Victorian reproductions of earlier styles are now

LEFT: An Ormolu-mounted Sèvres porcelain vase, 18th–19th century.

regarded as antiques in their own right, there will obviously be quite a wide difference in the antiquarian value. This is also the case with early 20th century reproductions of Louis Quinze. In porcelain it is often a greater problem especially where such factories as Meissen, Derby or Worcester revive old patterns. Generally speaking, however, variations in marking help to distinguish between the originals and the revivals. In all such cases of doubt, it is recommended that the would-be purchaser get the advice of a reputable dealer or auctioneer. Legislation in many countries in recent years, such as the Trades' Descriptions Act in Britain, place a grave responsibility on the vendors and their agents to ensure that articles are properly described.

At the end of the day, the age-old maxim *caveat emptor* is as important as ever; but do not let this deter you from enjoying the quest for your chosen subject. All collectors make mistakes along the way, but so long as they learn from the experience no great harm is done.

Delftware, Maiolica and Faience

Although a coarse earthenware was produced in the Dutch town of Delft, it actually originated in the Near East and its development may be traced to the western Mediterranean. Maiolica derives its name from Mallorca or Majorca, one of the Balearic Islands, and was used to describe the vigorously colored tin-glazed earthenware introduced to western Europe via the Hispano-Mauresque wares of Spain. It should not be confused with majolica, a term devised in the early 19th century to describe a type of earthenware with a thick, colored glaze, which was very popular for tiles in Victorian times. Maiolica is simply the name under which a popular type of tin-glazed enamel pottery was known in Italy, where it was produced in the Faenza district—hence the name faience by which it is known in France and Germany. Urbino was a notable center for Istoriato ware, plates, beakers and plaques decorated with historic scenes. Both Italy and Spain produced albarelli, cylindrical drug pots of a type that originated in medieval Persia. Maiolica was the forerunner, if not a parallel development, of the delftware popular in the Low Countries and England. It differs from delftware primarily in its brilliant coloring, and in subject matter, which favored religious, historical or mythological subjects.

Long before Delft established its international reputation in the 17th century, this type of pottery was being made in England. Henry VIII encouraged the immigration of Flemish potters, two of whom, Jasper Andries and Jacob Janson (whose name was eventually anglicized to Johnson) set up business in Norwich. Later Jacob Johnson established a pottery at Aldgate in London and by the end of the 16th century several Flemish potters were plying their trade in the Southwark and Lambeth districts.

Early English delftware of established provenance is extremely rare, the earliest examples being the famous Malling jugs in the British Museum and the Victoria and Albert Museum. In the 17th century, potteries were established in Bristol and the neighboring village of Brislington, while the London potters spread the technique to Liverpool. Factories of lesser importance flourished at Wincanton in Devon, at Limerick and Dublin and latterly in Glasgow.

True delftware may be recognized by its opaque glaze, usually decorated in a vigorous manner of short, sharp strokes, necessitated by the painter having to work on a porous surface, which very quickly absorbed the pigment. Experts distinguish the products of the various districts by the slightly colored tinge of the glaze or shades of pigment, which one pottery favored over another.

The most popular forms of delftware are the large "blue dash" chargers with their crude but vigorous portraits of royalty and biblical figures. The medium was also widely used for electioneering plates in the mid-18th century, and these now attract a keen following. Dutch delftware became much more sophisticated in the 18th century, and blue and white plaques often had attractive Rococo gilt frames.

Oriental Porcelain

A white porcelain, known as *blanc de Chine*, was developed at Te-Hua during the Ming dynasty and varies from a mellow creamy color to a hard bluish white. It was invariably used for the production of vases, which were either left undecorated or were painted in enamel colors for European consumption.

The rose, green, black and yellow families (*famille rose*, *verte*, *noire* and *jaune*) are the somewhat misleading terms used to denote groups of Chinese porcelain, distinguishable by their transparent enamel glaze. The earliest of these was *famille rose*, produced during the Ming period and recognizable by its rosy tinge and slight "orange-peel" texture. Genuine pieces of *famille rose* ware are now exceedingly rare. Porcelain production declined after the overthrow of the Ming dynasty in 1644, but during the reign of Kangxi (1662–1720) the imperial porcelain factory at Ching-te Chen began producing *famille verte* ware. The green family includes two shades of green, a coral pink, a bright yellow and a distinctive aubergine shade of purple. The glaze itself has a bluish tinge.

When the enamel was applied to black-outlined designs painted direct on the unglazed porcelain the result was known as *famille noire*. This technique in green, yellow or aubergine enamels gave a startling result and led the way to the gorgeous deep greens and greenish blacks that distinguish this family. Ornaments such as birds (usually parrots), human figures, lions, pagodas and horses were produced during the reign of Kangxi and his successors. Small sweetmeat dishes with attractively enameled animals and flowers are still fairly plentiful.

By far the commonest form of decoration on early Chinese wares consisted of blue on a white ground, the reason being that the blue pigment was the only one available to the early potters that was able to withstand the terrific heat of the kiln during firing. The willow pattern was only one of many decorating Chinese wares, but it was by far the most popular in the wares exported to Europe and was widely imitated by the European potteries from the mid-18th century onwards. During the 19th century, China developed a considerable trade in export porcelain, which was shipped to Europe and America through Canton. Canton ware was deliberately manufactured to suit Victorian European taste and is sometimes regarded as over-ornamented.

The art of porcelain did not spread to Japan until well into the 17th century, introduced by Korean immigrants in the Kutani area. Suitable clays were discovered in southern Kyushu in 1616, the year in which the potter Kakiemon settled in Arita. He is alleged to have discovered the secret of the red overglaze in 1640, although it is more probable that his son of the same name hit upon it between 1660 and 1690.

Kakiemon wares are remarkable and highly distinctive, recognizable by the delicate colors on that flawless milky white porcelain which the Japanese call nigoshide. Later wares extended the range of glazes and decorative treatments enormously. Although the bulk of production consisted of tablewares, the Japanese potters also produced various genre figures and candlesticks.

European China

The first true European porcelain was produced at Meissen in Saxony where Augustus the Strong established a pottery in 1710. The substitute for the costly Oriental porcelain was discovered almost accidentally by Johann Friedrich Böttger, though he then spent 11 years perfecting a material that resembled the hard, translucent porcelain invented by the Chinese.

The best period of Meissen was from 1720 until 1750 when J.J. Kaendler produced his delightful harlequinade figures. Meissen also specialized in porcelain flowers, but the bulk of production consisted of tableware. Meissen porcelain of the 19th and early 20th centuries has risen sharply in value in recent years.

Dresden china is a term applied to the products of other factories in Saxony, which often emulated Meissen. Other German areas whose output has long been appreciated include Nymphenburg (1754–64), Frankenthal (1755–99), Ludwigsburg (1758–1824), Berlin (from 1763) and Vienna (1744–1864).

All of them produced both tableware and ornamental pieces. Berlin also developed lithophanes, plaques which when held up to the light revealed a hidden image, especially popular in lampshades. Inspired by the example of Vienna and Saxony, many porcelain factories sprang up in Prague and other towns of Bohemia (now the western part of the Czech Republic). Royal Dux is perhaps the best known.

A porcelain factory was opened at Vincennes in 1738, and moved to Sèvres in 1756 where it continues to this day. It is particularly noted for its tableware, often beautifully hand-painted with genre scenes. Limoges and Paris are the other main centers of porcelain manufacture in France.

The Royal Copenhagen factory was established in 1755 and continues to this day. It is noted for its greyish blue glazes in both tableware and ornamental pieces. The private company of Bing and Grondahl has been in operation for 150 years, and likewise favors blue coloring. Both factories have long been noted for their figurines and commemorative rack-plates.

As early as 1744 there was an attempt to establish a porcelain factory in St Petersburg, but it was not until 1758 that production of wares modeled on those of Sèvres got under way. At its best, St Petersburg porcelain of the late-18th and early 19th centuries is of a very high standard, both in the quality of the paste and the firing and glazing. It continued under the patronage of the Tsar until 1917, and thereafter operated as a state-owned enterprise, but private factories also flourished—notably those run by Kornilov and Kuznetsov, both of which produced excellent genre figurines and groups. The Gardner family from England operated a factory in Moscow from 1758, its wares drawing heavily on those of Meissen, Sèvres and Vienna for inspiration, though after 1812 Gardner wares developed along highly individual lines. The Popov factory (1806–72) specialized in tableware decorated with realistic views of Russian scenery.

RIGHT: A massive 18th-century Chinese blue and white basin painted at the center with a river landscape scene.

INTRODUCTION

British Pottery and Porcelain

The earliest porcelain factory in England was founded by Nicholas Sprimmons at Chelsea in 1744. Among the earliest products were the famous "goat and bee" jug and the entrancing "girl on a swing" figures. The phases of the Chelsea factory are usually divided according to the marks: triangle (1745–49), raised anchor (1750–53), red anchor (1753–58), and gold anchor (1758–70). The work of the raised and red anchor periods are the most highly rated. The factory was purchased by William Duesbury and John Heath of Derby, and from then until its closure in 1784 its products closely resembled those of Derby. The factory, opened by Thomas Frye and Edward Heylyn at Bow in 1744, was in production by 1746, and survived until 1776. Its wares were characterized by their heavy, opaque appearance. It is credited with introducing transfer printing to porcelain, but is best remembered for its figures, second only in quality to Chelsea.

The manufacture of porcelain was certainly being carried on at Derby by 1751 if not before. William Duesbury took over a number of other potteries and built up a flourishing business specializing in Chinese figure groups as well as hollow wares. The original pottery closed in 1848, but it was resurrected in 1876 as the Royal Crown Derby Porcelain Company, noted for its ornamental wares and dinner services. The Royal Worcester Porcelain Company, founded in 1751 by Dr John Wall, is the oldest established pottery of its kind in Britain still in production. The company passed successively through the hands of the Flights and the Barrs in various combinations (1783–1840), the Chamberlains (1840–52), Kerr and Binns (1852–62) and has been known by its present name since the latter date. Other early potteries, whose porcelain is esteemed today, include Caughley (1775–99), Coalport (1795–present), Davenport (1793–1887), Minton (1793–present), Rockingham (1746–842), and Spode-Copeland (1770–present). Wedgwood (1759–present) produced fine porcelain from the early 19th century onwards, but is best known for its basaltes and jasperware. Similarly, Mason's porcelain factory (1796–present) diversified from porcelain into the polychrome ironstone for which it is best known. Doulton (1815–present) built its business on stoneware before diversifying into art wares, notably the character mugs and jugs, which continue to this day. Poole (1922–present) also produces stoneware including polychrome dishes and hollow wares.

Outside England, the chief names to note are Belleek in Northern Ireland, Wemyss in Scotland, and Nantgarw and Swansea in Wales, both of which flourished in the early 19th century and produced excellent floral decoration on tableware.

The Staffordshire potteries mainly produced utilitarian tableware, but also excelled in the manufacture of flat-back figures designed to decorate the mantelpiece. The best-known examples of this genre were the pottery figures of spaniels, but cows and other farmyard animals, milkmaids, sailors and highlanders were also popular. In the 19th century, figures of royalty and contemporary celebrities were produced for the mass market.

Art and Studio Pottery

Ceramics produced by individual craftsmen-potters, as opposed to the mass-produced wares of the great commercial potteries, developed out of the art pottery of the 19th century. Reacting against the lack of creativity or individuality of the leading European factories, some potters decided to turn their backs on useful wares and even the standard repertoire of decorative

INTRODUCTION

wares, and tried to create something that reflected the intellectual "esthetic" approach of the period.

Art pottery had many origins in many countries. Henry Cole, under the alias of Felix Summerly, pioneered art pottery as early as 1846, but with little success. Theodore Deck opened his studio pottery in Paris in 1856, and won wide acclaim at the Paris Exhibition of 1867. Thereafter the vogue for studio pottery was well established and spread rapidly to other parts of Europe.

The French studio pottery attained its peak at the turn of the century in the work of Dammouse, Dalpayrat and Delaherche; all concerned with the effects of various glazes and their elaboration on pottery and porcelain.

In England, Minton revived Cole's ideas in 1870, establishing the Art Pottery Studio where artists were encouraged to decorate Minton wares. The studio burned down in 1875 and was never revived, but it triggered off a number of other studios. The outstanding figures of this movement were William de Morgan, first at Merton Abbey and latterly at Sand's End, Harold Rathbone, who founded the Della Robbia Pottery at Birkenhead, and the Martin Brothers at Fulham, who specialized in grotesque stoneware figures of animals and birds and gave their name to Martinware. William Moorcroft (1913–present), produced distinctive art wares. The commercial factories did not entirely disregard the vogue for studio pottery. Doulton took up where Minton left off, and encouraged such artist-potters as Hannah and Florence Barlow. In more recent times works of individual designers, such as Clarice Cliff at Wilkinson and Susie Cooper at Wedgwood, are now much sought after.

Studio pottery developed rapidly in the Low Countries from 1885 onwards inspired by such diverse influences as early delftware, the English Arts and Crafts Movement and Javanese batik patterns. The Rookwood Pottery trained many of the American artist-potters, such as Laura Fry and Artus van Briggle, who branched out in the 1890s and later on. Samuel Weller's Louwelsa and Roseville potteries developed slip-painting. Other firms included Lonhuda, Grueby Faience, and the Dedham Pottery.

Studio pottery attained its highest level after 1920, the year in which Bernard Leach returned to England from Japan and established a studio at St Ives, working with Shoji Hamada. Michael Cardew (1901–83) worked mainly in slipware, though he also produced earthenware and stoneware. Lucie Rie (1902–95) and Hans Coper (1920–81) joined forces in 1946 and brought a distinctive Continental influence to bear on postwar studio pottery. Elizabeth Fritsche is the outstanding figure in contemporary studio pottery, working in colored clays and slips, and relying on decorative effects rather than sculptural forms.

Deruta Dish
c.1525

At a Glance

Date: c. 1525
Origin: Italy
Brief description: A Deruta bella donna dish.

The term maiolica was coined in Italy in the 15th century initially to describe lustered wares from Valencia in Spain, which were imported aboard sailing ships from Majorca (Mallorca). Soon the term was extended to Italian tin-glazed earthenware which imitated it. Although the chief centers of this industry were Faenza, Orvieto and Siena, Deruta near Perugia began by imitating Sienese wares before developing its own distinctive style, such as this heavy broad-rimmed dish depicting a bella donna with accompanying text. The inscribed "Ofacie soculis isidiosa meis" loosely translates as "o face insidious to my eyes."

DELFTWARE, MAIOLICA AND FAIENCE

Istoriato Dish
1533

At a Glance

Date: 1533
Origin: Italy
Brief description: An Istoriato (story-telling) dish from the workshop of Francesco Xanto Avelli da Rovigo.

The motif shows Europa and her companions admiring three bulls. In the background Zeus disguised as a white bull rides off with Europa. The back of the plate bears a caption to the picture as well as details of manufacture.

Pesaro Maiolica Istoriato Dish
c.1560

At a Glance

Date: c.1560
Origin: Italy
Brief description:
Pesaro maiolica Istoriato dish, which is possibly from the Zenobia workshop.

The saucer dish has a painting of Tobias wrestling with the Angel (identified on the underside). The well-known Biblical story is brought to life with humorous and homely touches, such as the little dog drinking water beside his master.

The scene of Tobias wrestling with the Angel was in fact based on a woodcut by Salomon Bernard, illustrated in Jean de Tourne's Figure del Vecchio Testamento con Versi Toscani *by Daniel Maraffi (Lyons, 1554).*

DELFTWARE, MAIOLICA AND FAIENCE

Urbino and Deruta Wares
c.1540–50s

At a Glance

Date: c.1540–50s
Origin: Italy
Brief description: Castel Durante tondino (c.1540), Deruta tondo (1554), Castel Durante dish (1555).

Production of maiolica spread in the 16th century to the Italian duchy of Urbino, and a pottery specializing in enameled earthenware was established at Castel Durante (now Urbania). Such master craftsmen as Zoua Maria and Nicola Pellipario established the reputation of Urbino for extravagantly decorated wares of all kinds, from tondini (coasters) and chargers to urns, fountains, and wine-cisterns.

Famiglia Gotica Albarello
late 15th century

At a Glance

Origin: Italy
Date: Late 15th century
Brief description: Jar for preserving drugs with Gothic script around waist that reads "Se aiabino."

Famiglia Gotica (Gothic family) is the Italian term for maiolica decorated in the medieval manner with inscriptions in heavy Germanic black lettering.

This Famiglia Gotica slender waisted albarello has blue Gothic script on an ocher and blue-lined manganese scroll flanked by scrolling foliage and peacock feathers, beneath the sloping shoulder with blue wave ornament. The interior has a lead glaze covering the bistre-colored body. It is probably from Faenza or Naples, late 15th century.

Castelli Albarello
c. 1540

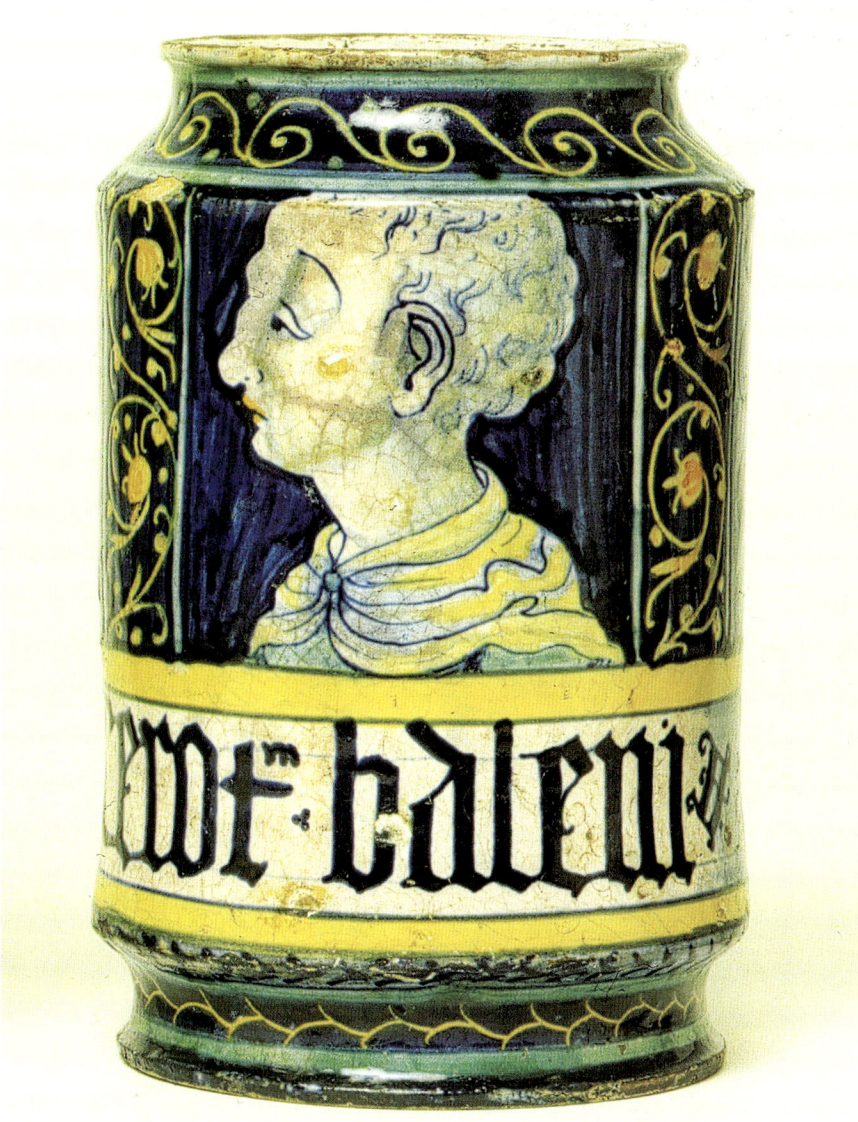

At a Glance

Date: c.1540
Origin: Italy
Brief description: A Castelli albarello of the Orsini Colonna type.

This albarello is painted with a male portrait in a rectangular cartouche between panels of yellow foliage scrolls above a yellow ribbon cartouche inscribed with the name of the contents "Cerot balem," the reverse with blue ribbons and stylized foliage. There is a blue band with yellow scale-pattern to the foot and the neck has a similar blue band with Vitruvian scrolls.

Spirit Flask
1685

At a Glance

Date: 1685
Origin: Germany
Brief description:
A barrel-shaped spirit flask.

Three scenes from the life of Christ appear in the foreground of a continuous landscape on this spirit flask painted in manganese, green, blue and ocher.

The date is inscribed on the underside. Frankfurt was a major center for the production of blue and white or polychrome faience.

DELFTWARE, MAIOLICA AND FAIENCE

Castelli Plaques
c.1720

At a Glance

Date: c.1720
Origin: Italy
Brief description: A pair of Castelli rectangular plaques.

The plaque on the right shows the Judgement of Solomon; the king is shown seated on a canopied throne surrounded by soldiers and courtiers. The other shows Joseph being sold into slavery by his brothers. The Ishmaelites are seen in the background mounted on their camels, and riches are heaped before the well.
  These plaques have yellow and black line borders and are painted in the colors of the Carlo Antonio Grue workshop. Earthenware plaques of this type were a speciality of Antonio Carlo Grue and his family at Castelli, who carried on the tradition of decorated maiolica till the mid-18th century.

Pesaro Maiolica Dish
early 18th century

At a Glance

Date: Early 18th century
Origin: Pesaro, Italy
Brief description: A large maiolica dish.

Although this piece is unmarked, the very high quality of the polychrome painting of the bouquet of flowers suggests the work of one of the great masters of this genre.

German Faience and Dutch Delft Group
18th century

At a Glance

Date: 18th century
Origin: Germany/Holland
Brief description: *(From left)* An armorial blue and white cream-pot from Brunswick or Osnabruck, c.1730; a Bayreuth oviform vase, c.1730; a pair of shoes; a Dutch candlestick, c.1700; a Dutch *famille verte* figure of an Oriental musician; and a bouquetier Oriental figure, c.1730.

Maiolica is known as faience in France and Germany because this type of tin-glazed enamel pottery was produced in the Faenza district of Italy. Maiolica was the forerunner, if not a parallel development, of the delftware popular in the Low Countries and England.

Dutch Delft Jars and Bottles
late 17th century

At a Glance

Date: Late 17th century
Origin: Holland
Brief description: A group of Dutch Delft apothecary jars and bottles with Latin names.

Apart from the very pleasing decoration, which includes cherubs, peacocks and floral garlands, these pieces are of immense interest on account of the Latin names of drugs and potions that date from a period when pharmacy was still virtually in the realms of witchcraft.

DELFTWARE, MAIOLICA AND FAIENCE

Dutch Delft Vases
c.1700

> **At a Glance**
>
> **Date:** c.1700
> **Origin:** Holland
> **Brief description:** Three Dutch Delft blue and white vases.

These vases bear the mark of Pieter Gerritsz Kam at de 3 Vergulde Astonnekens. They have been decorated in the Persian style, with birds amid flowers above a band of stylized leaves, the shoulders with roundels bearing flowers on a white ground alternating flowers on a blue ground. The larger ovoid vase with flat cover is flanked by two small bottle-shaped vases.

ANTIQUES AT A GLANCE: CERAMICS

English Delftware Group
late 17th /early 18th century

At a Glance

Date: Late 17th/early 18th century
Origin: England
Brief description: A group of English delftware, comprising a wet-drug pot, a charger depicting a tower and a dish with an Oriental scene, both probably from Brislington about 1680–90, and a puzzle fuddling-cup of six globular cups from Bristol or London, about 1740.

In the 17th century delftware potteries were established in Bristol and Brislington, while the London potters spread the technique to Liverpool. Delftware can be recognized by its opaque glaze, usually decorated in short, sharp strokes. Experts distinguish products from various districts by the slightly colored tinge of the glaze or shades of pigment.

DELFTWARE, MAIOLICA AND FAIENCE

English Delftware Group
c.1680

At a Glance

Date: c.1680
Origin: London
Brief description: A group of English Delft hollow wares.

Covered jars with elaborate tops are seen at the rear along with an elegant vase. The bowls in the front flank a posset pot of globular form with everted mouth, scroll handles and spout, and painted with Chinese figures in a landscape.

English Delftware Group
early 18th century

At a Glance

Date: Early 18th century
Origin: England
Brief description: A selection of blue and white delftware.

The plate with scalloped form was very fashionable at this time and continued to influence dishes in other materials well into the century. As well as the three covered pots there is a fine example of a posset pot (lower right) on a tripod base, painted with a boat, house on islands and sprays of flowers in the Chinese style which was all the rage at the time. For good measure the finial on the cover is modeled in the form of a recumbent Chinaman.

DELFTWARE, MAIOLICA AND FAIENCE

Henry Delamain Wares
c.1755

At a Glance

Date: c.1755
Origin: Ireland
Brief description: A group of wares from Henry Delamain's factory.

Dublin was an important center for the manufacture of delftware, both in the traditional blue and white, and polychrome glazes. In addition to a selection of dishes and a mug the picture shows a rare example of a basket whose sides are pierced with interlocking circles painted with a cell pattern and stylized garlands and pendants. The interior of the basket is painted with a vignette of a figure carrying a staff in a rock-strewn landscape, with birds in flight over fields in the background.

ANTIQUES AT A GLANCE: CERAMICS

Dublin Delftware Plates
c.1760

These plates show the gradual evolution of decorative styles and a more adventurous use of colors other than the traditional blue. The small plate on the left has an artless nautical motif of a sailing boat, the rim decorated in contrasting colors of pale blue and gold.

The larger plate on the right, however, is painted in manganese with two figures beneath a tree on prominent rockwork, in a shrubby mountainous but stylized landscape, within a border of four groups of foliage beneath drapery swags.

At a Glance

Date: c.1760
Origin: Ireland
Brief description: Two Dublin delftware plates by Henry Delamain.

DELFTWARE, MAIOLICA AND FAIENCE

Nuremberg Plaque
c.1740

At a Glance

Date: c.1740
Origin: Germany
Brief description:
A Nuremberg oval plaque painted in *grand feu* colors.

The plaque shows Solomon and another figure. Biblical subjects were a particular favorite with German faience decorators. This plaque was pierced so that it could be hung on a wall.

ANTIQUES AT A GLANCE: CERAMICS

French Faience Tureen
mid 18th century

At a Glance

Date: Mid 18th century
Origin: Lille, France
Brief description: A French faience chicken tureen and cover.

Faience in novelty shapes was a speciality of the French factories. This tureen and cover is naturalistically modeled in the shape of a hen and chicken with polychrome decoration. The marks on the cover and base indicate that it was produced at Lille in the middle of the 18th century.

ORIENTAL PORCELAIN

Kakiemon Dragon
late 17th century

At a Glance

Date: Late 17th century
Origin: Japan
Brief description: A fine and rare Kakiemon model of a dragon decorated in iron-red, blue, green and yellow enamels on a white glazed body.

The dragon's scaled tail is coiled tightly to one side, its head peering straight upwards with mouth agape revealing jagged teeth. The tama is held in its claws and tucked under its chin beside a distended stomach.

Imari Vase
late 17th century

At a Glance

Date: Late 17th century
Origin: Japan
Brief description: An Imari baluster vase and cover.

This vase is of unusual design and may possibly have followed a Delft design which in turn had interpreted a Japanese design, for example that of the delft-wares of Pieter Adriaenz Kocks.

It is decorated in iron-red and black enamel and gilt on underglaze blue to show a mountainous landscape with houses, trees and clouds. The cover is similarly decorated and the finial modeled as a bird of prey.

ORIENTAL PORCELAIN

Imari Dish
c.1700–10

At a Glance

Date: c. 1700–10
Origin: China
Brief description: An Imari armorial octagonal dish.

This massive dish is believed to have been made for the Corbeau, Corbel or Corbet families of Brittany or Normandy (all of whom have three crows in their arms, although there is also a possibility that it was made for a Dutch family). The border design closely follows its Japanese original, which was also copied at the Meissen factory. This is a good example of the decorative style of one country being copied by a second for sale in a third and reflects the eclecticism of Oriental art for the European market.

Cloisonné Vase
late 19th century

At a Glance

Date: Late 19th century
Origin: Japan
Brief description: A large cloisonné enamel oviform vase decorated in iris and water plants on a dark blue background.

This form of enameling takes its name from the cloisons or tiny cells composed of wire soldered in honeycomb fashion to the surface of the object to be decorated. The transparent enamel is then poured into the cells and allowed to harden. This technique was widespread in medieval Europe but then declined. It remained popular in China and Japan, whence it was reintroduced into Europe enjoying immense popularity from 1880 to 1900 when the fashion for japonaiserie *was at its height.*

ORIENTAL · PORCELAIN

Cloisonné Vase
c.1900

At a Glance

Date: c.1900
Origin: Japan
Brief description: A large cloisonné enamel vase.

The vase is in the baluster form decorated in colored enamels and with shakudo *fittings with a musen design depicting flowering cherry trees in the mist. The base has a gilt lacquer inscription: "Jamusu-kun ni okuru Meiji sanjuyon-nen jugatsu sakuma kozaburo," which translates as "Presented to Mr. James by Sakuma Kozaburo, Meigi 34 (1901) 10th month."*

Famille Rose Vase
19th century

At a Glance

Date: 19th century
Origin: China
Brief description: A *famille rose* turquoise ground molded rouleau vase.

The sides have high-relief molded vases, jardinières, scholars receptacles, teapots and ewers variously containing fruiting and flowering branches among other Buddhist and Daoist emblems.

ORIENTAL PORCELAIN

Famille Rose Vase
1875–1908

At a Glance

Date: 1875–1908
Origin: China
Brief description: A *famille rose* "one hundred bats" bottle vase with a Guangxu six-character mark.

About 1726 a new palette of colors was introduced and spread rapidly in the early years of Qianlong's reign. To the Chinese these colors were known as juan is ai (literally "soft colors"). The French term alludes to the predominantly rose or pink shade of the glaze derived from powdered gold. As it displaced the earlier famille verte, *these new pigments were adapted to a more effeminate style of decoration, and minuscule motifs noted for their delicacy.*

The vase is finely enameled with iron-red bats and colorful clouds between gold bands, petal panels, and a lotus frieze with shou characters around the shoulder. The "hundred bats" was one of the newly composed designs of the reign.

ANTIQUES AT A GLANCE: CERAMICS

Famille Rose Bowl
Yongzheng period

At a Glance

Date: Yongzheng period
Origin: China
Brief description: A superb *famille rose* coral-ground bowl that has a four character mark within a double square.

The rounded sides of the bowl are exquisitely enameled in shades of pink, blue, white, green, yellow and iron-red against a rich coral ground with lush pink poppy blossoms and buds among large curly leaves. Only one other Yongzheng famille rose bowl with this motif has been recorded (in the celebrated Avery Brundage Collection). The style and decoration of this bowl show a level of sophistication that suggests production late in the Yongzheng period.

ORIENTAL PORCELAIN

Famille Rose Hollow Wares
c.1800

At a Glance

Date: c.1800
Origin: China
Brief description: A group of *famille rose* hollow wares showing a variety of vases, with or without handles and covers, as well as a porcelain box with curved and domed cover and side handles (*below center*).

The two slender waisted baluster vases flanking the box have been mounted as lamps. The group is characterized by vignettes of ladies and gentlemen, pavilions, scenery and flowers, typical of the wares produced around 1800 and destined mainly for the European market.

ANTIQUES AT A GLANCE: CERAMICS

Famille Rose Wares
Qianlong period

At a Glance

Date: Qianlong period
Origin: China
Brief description: *Famille rose* wares including figures, lamp, vases and a watch-holder.

The small cylindrical vases were used as spill-holders. The watch-holder and cover of shaped oval outline is shown on the far left, and has a fluted base and an overall decoration of blooming vines.

ORIENTAL PORCELAIN

Famille Verte Elephant and Figures
19th century

At a Glance

Date: 19th century
Origin: China
Brief description: A pair of *famille verte* elephant and immortal figures.

Porcelain of this classification derives its French name, famille verte, *(green family) from the predominance of this color in many different shades. Each of the figures is modeled standing foursquare and turning the head to one side with the trunk held up. The Immortals are mounted side-saddle with one leg pendant and the other one raised. They hold a scholar's taper stick and wear an arrogant expression as befits their exalted status.*

55

Yuan Jar
14th century

At a Glance

Date: 14th century
Origin: China
Brief description:
A magnificent and rare Yuan blue and white *Guan* (jar).

The jar has a very large ovoid shape with flanged shoulders applied with a pair of dragon-fish handles. The body is decorated with two striding four-clawed dragons. This vessel appears to be unique in its decoration, although it can be linked to other examples of the Yuan dynasty, either excavated or from important collections.

ORIENTAL PORCELAIN

Dragon Jar
Jiajing period

At a Glance

Date: Jiajing period
Origin: China
Brief description: A large blue and white dragon jar with a Jiajing six-character mark.

Heavily potted and painted in deep tones of underglaze blue, the jar features a pair of scaly, five-clawed dragons racing through clouds above a band of waves and rocks, the two separated by shou *characters formed by twisting stems ascending from lingzhi fungus sprigs growing from the rocks.*

The motifs favored in this period reflect the influence of Daoism, although elements from earlier periods, such as the dragon, phoenix, flower scrolls and children at play continued as well.

ANTIQUES AT A GLANCE: CERAMICS

Yaoling Zun
Kangxi period

At a Glance

Date: Kangxi period
Origin: China
Brief description: Mallet-form vase bearing the six-character reign mark of Kangxi (1662–1723).

Most of the Chinese porcelain that comes on the market belongs to the period of the Manchu dynasty, which ruled from 1644 to 1911. It is broken down into the products of individual reigns, of which the most brilliant were those spanning the years from 1662 to 1795.

This elegant form is known as yaoling zun *(literally "vase in the shape of a handbell")*. Its unusual shape originated in the classic paper-beater vases, zhichui ping of the Song period. The minimalist ornament and large white areas are only found in porcelain of this reign.

ORIENTAL PORCELAIN

Famille Verte Stembowl
Kangxi period

At a Glance

Date: Kangxi period
Origin: China
Brief description: A *famille verte* stembowl and cover.

The curved sides spread to a straight rim, thinly potted and supported on a tall, slender, hollow foot, the cover domed with a flaring rim. The finial is molded in relief with gnarled branches in openwork extending to a field of painted fruiting peach branches, while the exterior of the bowl is ornamented with a continuous band of peach branches.

Famille Verte "Birthday" Dish
1713

At a Glance

Date: 1713
Origin: China
Brief description: A rare *famille verte* "birthday" dish with the six-character reign mark of Kangxi. The dish is delicately enameled with a central motif of a bird perched on a fruiting peach branch, the everted rim richly decorated in iron-red with lotus florets on a honeycomb diaper ground.

The nickname for the dish comes from the fact that such dishes bore highly auspicious decorations and inscriptions representing imperial birthday greetings. This dish is believed to have been produced to mark the 60th birthday of the Emperor Kangxi in 1713. These dishes were presented to courtiers and officials to be preserved as mementoes of a great occasion.

ORIENTAL PORCELAIN

Famille Verte Wares
early 18th century

At a Glance

Date: Early 18th century
Origin: China
Brief description: A group of *famille verte* wares. As well as dishes and covered bowls there is an unusual rectangular vase (far left) fitted with gilt metal mount and liner.

These wares are characterized by their extravagant decoration in which fruiting peach branches and flowering prunus were popular and recurring motifs.

ANTIQUES AT A GLANCE: CERAMICS

Famille Verte Group
early 18th century

At a Glance

Date: Early 18th century
Origin: China
Brief description: Group of *famille verte* figurines, the majority of which are joss-stick holders.

Many of the small figurines, on closer inspection, turn out to have had some useful purpose. The majority of the items in this picture, for example, are pairs of figures serving as joss-stick holders, modeled as Buddhist lions or as figures of Immortals.

ORIENTAL PORCELAIN

Famille Verte Vases
18th/19th centuries

At a Glance

Date: 18th/19th centuries
Origin: China
Brief description: A pair of *famille verte* vases of square tapering form, one decorated with courtly figures and warriors, the other with landscapes and dragons.

These porcelain vases date from the Kangxi period, but the ormolu mounts were added in the 19th century. The channeled scrolling ormolu mounts on these vases are related to a pair of rouleau vases formerly in the J. Paul Getty Collection. Similar ormolu mounts have also been recorded on Berlin porcelain of the 19th century, leading to the supposition that the mounts were German in origin.

Ming-style "Lotus Bouquet" Dish
1723–35

> **At a Glance**
>
> **Date:** 1723–35
> **Origin:** China
> **Brief description:** Blue and white Ming-style "lotus bouquet" dish with a Yongzheng six-character mark in a double circle.

Typical of the Chinese ceramics produced for export to the West is this handsome blue and white Ming-style "lotus bouquet" dish. It is dated by the Yongzheng six-character mark in a double circle, although the central motif has been recorded on dishes from the early 15th century, characterized by simulated "heaping and piling" with a continuous composite floral meander between the key-pattern around the rim and classic scroll around the base.

Ming-style Moonflask
c.1725–35

At a Glance

Date: c.1725–35
Origin: China
Brief description: An exquisite blue and white Ming-style moonflask with a six-character mark of the Yongzheng period.

The flattened circular body is superbly painted on both sides in vibrant tones of blue in imitation of the "heap and pile" effect and an attractive array of different flowers. No other Yongzheng vase of this type has been recorded, although there are several other vases of this shape painted only with a single fruiting spray on each side. While the form and style of decoration on this flask are inspired by early 15th-century originals, no Ming flask of this pattern appears to have been recorded so far.

Ming-style Lotus Dish and Stemcup
Qianlong period

At a Glance

Date: Qianlong period
Origin: China
Brief description: A Ming-style blue and white lotus dish and a stemcup of a similar design with a Qianlong six-character reign mark of the period 1736–95.

The center of the dish features a ribboned lotus bouquet below a composite floral scroll in the well. The design on this dish is a close copy of a type occasionally found on Yongle dishes of a much earlier period.

The stem cup is in the early Ming style with lanca *characters above lotus flowers. Stem cups with this design reflect the Qianlong emperor's interest in Lamaist Buddhism, which flourished during his reign.*

ORIENTAL PORCELAIN

Chinese Porcelain
mid 18th century

At a Glance

Date: Mid 18th century
Origin: China
Brief description: An array of Chinese export porcelain recovered from the wreck of the East Indiaman *Nanking*.

Salvage operations in the 1980s brought to light a vast quantity of blue and white wares destined for the European market. This selection, retrieved from the Nanking, *reveals the range and diversity of the decorative styles, many dating from earlier periods, as well as the pictorial vignettes adapted to suit Western taste.*

Chinese Porcelain Vases
late 19th century

At a Glance

Date: Late 19th century
Origin: China
Brief description: A pair of porcelain blue and white vases.

The vases were subsequently fitted with gilt bronze mounts and adapted as table lamps. The baluster bodies and tall, waisted necks are decorated with pheasants amid trees and foliage.

ORIENTAL PORCELAIN

Baluster Jars and Covers
Kangxi period

At a Glance

Date: Kangxi period
Origin: China
Brief description: A pair of baluster vases.

Each jar has a painted body with two panels depicting ladies and boys on and around a castellated wall with domed turrets, and stylized flowerheads above stiff leaves. There are bands of demi-florettes at the shoulder and artemisia leaves around the base.

Meissen Foliate Dish
c.1735

At a Glance

Date: c.1735
Origin: Germany
Brief description: Meissen foliate dish with blue crossed swords and dot marks.

The dish is painted in the Japanese Imari style, the center with indianische Blumen, a German term meaning "Indian flowers," which was used to distinguish this style of Oriental floral decoration. The well and border have radiating and alternating shield-shaped and blue ground lappets with stylized flowerheads contained in a narrow blue-ground border with a waved gilt line. This is a splendid example of the rich, sumptuous colors that distinguish the soft paste porcelain of this period.

EUROPEAN CHINA

Meissen Porcelain Trifles
c.1730–35

At a Glance

Date: c.1730–35
Origin: Germany
Brief description: Objects of vertu, including snuffboxes, made from soft-paste porcelain with gold mounts of a later period.

The versatility of Meissen is revealed in this array of objects of vertu, the delicate soft-paste porcelain of the early period lending itself admirably to all manner of small objects such as snuffboxes, patch boxes, pounce-pots, scent bottles and spirit flasks.

The porcelain trifles in this picture were manufactured and decorated between 1730 and 1735, but in some cases the gold mounts bear much later assay marks.

ANTIQUES AT A GLANCE: CERAMICS

Meissen Porcelain Trifles
18th century

At a Glance

Date: 18th century
Origin: Germany
Brief description: A selection of Meissen porcelain trifles including a fine pipe bowl modeled as a human head, an etui (needle-case) and a pair of silver gilt-mounted snuffboxes.

The snuffboxes illustrate the misadventures of Don Quixote and also reflect the universal appeal of Cervantes' latter-day knight errant all over Europe in the 18th century.

EUROPEAN CHINA

Meissen Snuffbox
c.1760

At a Glance

Date: c.1760
Origin: Germany
Brief description: A Meissen silver-gilt mounted snuffbox of about 1760, although the mounts probably belong to the 19th century.

The box is formed as a naturally modeled shell painted with gray and pale puce markings and encrusted with smaller shells and crustaceans. The interior (not shown) is richly gilt, while the inside of the lid has a miniature finely stippled painting showing a couple of pilgrims (identified by the shells attached to their clothing), hand in hand before a distant galleon. This is the kind of romantic genre scene that found great favor in the snuffboxes and pillboxes of the 18th century.

Meissen Snuffboxes
c.1759–60

At a Glance

Date: c. 1759–60
Origin: Germany
Brief description: Three gilt-metal mounted Meissen snuffboxes.

These snuffboxes show the skill of the Hausmalerei in reproducing very detailed and intricate genre subjects and landscapes in such a confined medium. The interior of the lid of the box on the right, for example, shows gallants and their companions in a park, with a splendid view of the Albrechtsburg in the distance.

Meissen Armorial Dish
c.1737–41

At a Glance

Date: c.1737–41
Origin: Germany
Brief description: A Meissen armorial dish from the Swan Service modeled by J.J. Kaendler and J.F. Eberlein.

The dish has shell molding about a molded centre showing two swans swimming among bulrushes beside a heron, fish and shells. The border is decorated with scattered indianische Blumen (translated as "Indian flowers") and the arms of Heinrich Graf Bruehl, 1737–41.

ANTIQUES AT A GLANCE: CERAMICS

Böttger Hausmalerei Beaker
c.1730

At a Glance

Date: c. 1730
Origin: Breslau, Germany
Brief description: A Böttger Hausmalerei beaker painted by Ignaz Bottengruber.

The beaker shows Neptune and a seahorse. Johann Friedrich Böttger, employed by Augustus the Strong as an alchemist, stumbled across the secret of hard-paste porcelain and launched the fortunes of the Meissen factory in 1710.

EUROPEAN CHINA

Möllendorff Dinner Service
c.1761–63

At a Glance

Date: c.1761–63
Origin: Germany
Brief description: Möllendorff silver-gilt mounted porcelain dinner service.

Although Meissen is best remembered for its magnificent hollow wares and figurines, the Saxon factory also specialized in all kinds of minor porcelain, from the handles of cutlery to the coins used in Dresden during the economic crisis that hit Germany after World War I.

Dinner services not only comprised a full set of table wares, but also the handles of the knives and forks. The beautiful cutlery in this picture from the Möllendorff Service consists of porcelain made in 1761–63 from models of Frederick the Great of Prussia and Karl Jakob Christian Klipfel, but the silver-gilt mounts were assembled at London about 1937, to judge from the assay marks.

ANTIQUES AT A GLANCE: CERAMICS

Meissen Dinner, Dessert and Side Plates
c.1860

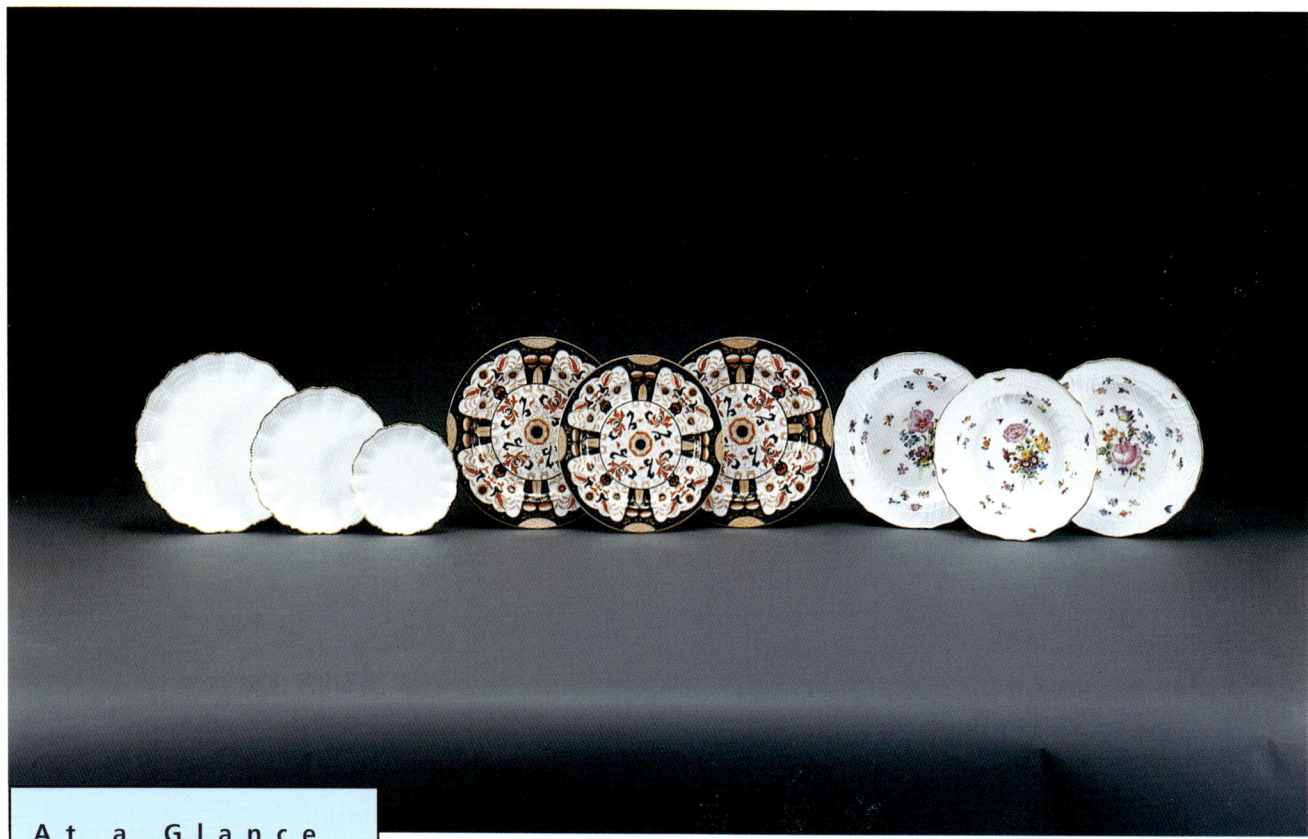

At a Glance

Date: c.1860
Origin: Germany
Brief description: Examples of the dinner, dessert and side plates from three different Meissen services.

These plates show the wide divergence in decorative treatment of the period, from the austerely simple style (left) in which the form rather than ornament is all-important, to the extravagantly decorated styles (center and right) with floral sprays and osier borders.

EUROPEAN CHINA

Four Meissen Figurines
late 19th century

At a Glance

Date: Late 19th century
Origin: Germany
Brief description: A group of four Meissen figurines, including a large serving dish and, in front, a two-handled oval vegetable tureen and domed cover.

The winning Meissen formula of the 1740s was still going strong a century and a half later. The finial on the domed cover of the vegetable tureen shows Meissen's uncanny skill in adapting the technique of the figurine to the handles of table ware; in this instance it takes the form of a putto spilling the contents of a cornucopia, painted with scattered flowers and insects.

ANTIQUES AT A GLANCE: CERAMICS

Meissen Part Dinner Service
late 19th century

At a glance

Date: Late 19th century
Origin: Germany
Brief description: A Meissen part dinner service painted with a loose spray of flowers and lesser sprigs, beneath a border painted with insects and scattered sprigs, all within a shaped gilt rim.

Although the full service would originally have been much larger, this group, consisting of soup plates, dinner plates, dessert plates, a fish plate, oval serving dishes in various sizes, two large circular plates and a mustard pot (not shown), is still sufficiently comprehensive as to find a ready sale.

EUROPEAN CHINA

Meissen Dishes
c.1880

At a Glance

Date: c.1880
Origin: Germany
Brief description: A group of Meissen circular dishes, comprising two soup plates and a circular dish.

The insects painted among a loose bouquet of flowers and scattered springs, are a form of decoration characteristic of Meissen tablewares for more than two centuries.

ANTIQUES AT A GLANCE: CERAMICS

Meissen Group
c.1886 onwards

At a Glance

Date: c.1886
Origin: Germany
Brief description: A Meissen group of three warriors and a Blackamoor on the back of an elephant.

Since Roman times the elephant has traditionally been associated with military victory. The elaborate howdah in the form of a turret was the original of the elephant and castle emblem used by the African Company.

EUROPEAN CHINA

Hofservice Dish
c.1760–65

At a Glance

Date: c. 1760–65
Origin: Bavaria, Germany
Brief description:
A Nymphenburg floral oval dish from the Hofservice.

Second only to Meissen in antiquity and reputation, the Nymphenburg factory in Bavaria was founded about 1753, and like Meissen it has long been renowned for its figurines and tablewares. This floral oval dish from the Hofservice of 1760–65 is decorated with gilt and colors, probably painted by Joseph Zächenberger, head of the Blumenmalerei (floral painting team) between 1760 and 1770. The service was originally produced for the Elector of Bavaria.

Scent Bottle
c.1790

At a Glance

Date: c.1790
Origin: Naples, Italy
Brief description: A scent bottle in the form of a Pierrot crouching on a tortoise.

Under Ferdinand IV, the royal porcelain factory at Naples produced exquisite porcelain figures in the late 18th century. This rather whimsical figure forms a scent bottle.

EUROPEAN CHINA

Naples Coffee Cup and Saucer
c.1790

At a Glance

Date: c.1790
Origin: Naples, Italy
Brief description: A Naples coffee cup and matching saucer bearing incised marks, which date them to about 1790.

The cup has a very distinctive key handle, and both cup and saucer are painted with friezes of classical warriors set against black ground panels flanked by medallions with doves below gilt swags of leaves.

The saucer has friezes of classical scenes and medallions with silhouettes of classical figures in the same manner. The decorative treatment reflects the craze for all things classical in Italy in the second half of the 18th century when the fashion for the Grand Tour was at its height, and porcelain wares of this type were a popular memento.

Doccia Snuffbox
c.1760

At a Glance

Date: c.1760
Origin: Italy
Brief description:
A Doccia gilt-metal mounted oval snuffbox.

This snuffbox was perhaps painted by Giovacchino Rigacci and is inscribed around the base with verses from Dante's Inferno. The quotations from Dante, though written many years before, describe the two groups of figures on the interior and exterior of the cover. The exterior of the cover has half-length portraits of figures from the Reformation including Luther, Calvin, Erasmus, John VI of Nassau and Renata, Duchess of Ferrara in black and purple robes. The interior has figures from the Counter-Reformation including Cardinals and theologians from the Catholic church.

EUROPEAN CHINA

Du Paquier Tableware
19th century

At a Glance

Date: 19th century
Origin: Vienna, Austria
Brief description: Tableware including a tankard with gilt-metal mounts.

The Vienna factory began in 1718 when Du Paquier experimented with porcelain, but in 1744 it was nationalized and continued under state management until its closure in 1864. Its sumptuous tablewares were noted for the richness of their colors, which, contrasted with gilt work, gave them a very exotic appearance. In pieces such as the tankard (top center) this effect was enhanced by the use of gilt-metal mounts.

ANTIQUES AT A GLANCE: CERAMICS

Vincennes Miniature Cabaret
c.1752–1806

At a Glance

Date: c.1752–1806
Origin: France
Brief description: Vincennes miniature cabaret comprising a milk jug, sugar bowl and cover, two-handled cup and saucer, and a triangular lobed tray.

A French royal porcelain works was established at Vincennes in 1738 and moved to Sèvres in 1756. This composite bleu lapis miniature cabaret dates from the last year of production at Vincennes and was painted by Etienne Evans who worked for the company from 1752 to 1806.

EUROPEAN CHINA

Sèvres Armorial Dessert-plate
1792

At a Glance

Date: 1792
Origin: France
Brief description: An armorial dessert-plate from the service made by Sèvres in 1792 for Henry Sudell of Box, Wiltshire.

Each plate was painted by J.N. Le Bel, copying F.N. Martinot's engravings of exotic birds (1771–86). Very few Sèvres armorial services exist and only a few pieces from the Sudell service have survived.

Sèvres Tablewares
c.1775

At a Glance

Date: c.1775
Origin: France
Brief description: A group of early Sèvres tablewares including a richly gilded teacup and saucer, and two plates.

The plate on the right has a lobed format, painted with a spray of roses within a gilt roundel, below a border painted with two leafy garlands entwining two gilt bands among loose sprays of flowers. The marks on the underside include the date letter for 1775, the painter's mark for Tandart and four blue dots on the foot rim.

EUROPEAN CHINA

Sèvres Plates
c. 1811–13

At a Glance

Date: c. 1811–13
Origin: Sèvres
Brief description: A set of 12 Sèvres blue-ground plates with various Sèvres printed marks deleted, and gilder's and incised marks substituted.

Sèvres was noted especially for the sumptuousness of its table services in which deep, dark colors contrasted with lavish gilding. In this instance a gilt central patera is surrounded by a broad blue border with gilt scrolling foliage and husk ornament.

Sèvres Serving Dish
1853

At a Glance

Date: 1853
Origin: France
Brief description: A Sèvres matt-red ground assiette plate (serving dish) from a service commissioned by Napoleon-Joseph-Charles-Paul Bonaparte.

The center is painted with a vignette of a dove perched on the side of a dish by an amphora above a pearl necklace. This strange motif is contained within fretted roundels while the border is gilt with trailing leaves. It has the printed iron-red crowned N mark and S.53 in green, indicating that it was made in 1853, the year following the purchaser's kinsman's seizure of power as the Emperor Napoleon III.

EUROPEAN CHINA

Dihl and Guerhard Plate
early 19th century

At a Glance

Date: Early 19th century
Origin: Paris, France
Brief description: Five Paris (Dihl and Guerhard) plates painted by Drölling with peasants at various pursuits.

At their Paris factory Messrs Dihl and Guerhard did much to raise the standards of design and quality of production of French porcelain at the beginning of the 19th century. They were particulary noted for their plates, employing the best artists, such as Martin Drölling (1752–1829) to paint genre scenes and landscapes.

ANTIQUES AT A GLANCE: CERAMICS

Gardner and Popov Porcelain Figures
19th century

At a Glance

Date: 19th century
Origin: Russia
Brief description: Two of the groups are by Gardner but the others were produced by the Popov factory and date between 1880 and 1917.

Porcelain manufacture in Tsarist Russia had a chequered career, beginning in 1744 with the unsuccessful experiments by Christoph Konrad Hunger. His St Petersburg factory was taken over by the State in 1758 and continues to this day, although its finest wares date from the late 18th century.

Best known today, and the one whose products are now most sought after was that founded at Moscow in 1758 by an Englishman, Francis Gardner. Several generations of the Gardner family operated the porcelain works at Verbilki, producing a fine range of tablewares. But what Gardner is best known for are the genre figures and groups of children and peasants. These endearing little figures date from about 1812 when a tax on tablewares forced Gardner to diversify. Gardner's success induced rival firms to emulate them.

EUROPEAN CHINA

Gardner Porcelain Figures
c.1812–20

At a Glance

Date: c.1812–20
Origin: Russia
Brief description: A selection of figures from the Gardner factory, including a milkmaid and balalaika player.

This further selection of porcelain figures from the Gardner factory dates from the earliest period (1812–20) and includes a balalaika player and a milkmaid, whose buckets are fitted with detachable covers. Gardner figures are marked with the letter G in Cyrillic.

Mettlach Plate
19th century

At a Glance

Date: 19th century
Origin: Luxembourg
Brief description: A Mettlach polychrome pottery plate.

Production of fine quality ceramics began in the Rhineland town of Mettlach in 1809 and over the ensuing decades the firm of J.F. Boch & Buschmann built up a reputation for their tin-glazed earthenwares. Polychrome plates were a specialty and ranged across the world in search of inspiration. The Japanese motifs on this polychrome pottery plate reflect the passion for the arts of this country after it was opened up to Western influence—a two-way traffic to judge by the craze for japonaiserie in the late 19th century.

EUROPEAN CHINA

Villeroy & Boch Vases
late 19th/early 20th century

At a Glance

Date: Late 19th/early 20th century
Origin: Luxembourg
Brief description: Three polychrome majolica vases, including a vase (*right*) depicting peasant girls.

The partnership of Villeroy & Boch dates from 1766 when Jean-François Boch established a factory at Septfontaines in the grand duchy of Luxembourg. The two families intermarried from 1836 onwards. A second factory under their joint names opened at Wallerfangen (Vaudrevange) in 1789 and later operations were also carried out at Mettlach (from 1842) and Dresden (from 1856). Originally producing white and cream-colored earthenwares, this firm began making decorative pieces in the late 19th century, specializing in polychrome majolica plates, vases and bowls with allegorical or genre subjects.

Clément Massier Jardinière
19th century

At a Glance

Date: 19th century
Origin: Golfe Juan, France
Brief description: A Clément Massier hand-painted ceramic jardinière.

The brothers Clément and Jérôme Massier established a studio pottery at Golfe Juan on the French Riviera in 1870. Both brothers produced a prolific range of vases, bowls and larger pieces, which were hand-painted, using combinations of pigments, enamels and gilding to produce some startling effects that were well ahead of their time.

They collaborated with Lucien Levy and other artist potters at the turn of the century as well as turning out pieces, which they individually signed, such as this jardinière by Clément. After his brother's death, Jérôme moved to Vallauris and re-opened his pottery, and over the ensuing years produced some of the early ceramic pieces that were painted by Picasso.

BRITISH POTTERY AND PORCELAIN

Staffordshire Slipware Jug
c.1700

At a Glance

Date: c.1700
Origin: England
Brief description:
A Staffordshire slipware jug and cover in the form of an owl.

This jug has the characteristic lead-glazed red earthenware slip-decorated in cream and brown which makes these pieces so readily identifiable. The ceramic liquid known as slip was poured from a container through a narrow spout which enabled the artist to "trail" it and thus form the raised decoration, in this case the intricate pattern of the bird's feathers, enhanced by a form of scratched comb decoration known as sgraffiato. Slipware was the earliest form of distinctive English earthenware; although unmarked, it was produced by a number of potters in the Burslem area, the most notable being Thomas Toft.

ANTIQUES AT A GLANCE: CERAMICS

Staffordshire Hawks
c.1755

At a Glance

Date: c.1755
Origin: England
Brief description: Staffordshire saltglaze models of a hawk on a rockwork base.

Both models were produced about 1755, the former richly enameled in bright colors and the latter in white. The exuberant colors (or lack of them) make a world of difference to the value of these figures.

Staffordshire Cow Creamers
19th century

At a Glance

Date: 19th century
Origin: England
Brief description: Polychrome earthenware cow creamers, including one in agateware (*left*).

Cow creamers were little milk jugs in the form of a cow, the tail curled to form the handle and the mouth forming the spout. They were inspired by silver cow jugs imported to England in the 1750s and were copied by all the potteries, usually in polychrome earthenware but often in distinctive materials and glazes, such as agateware (*left*), an unusual form of stoneware which derives its name from the veined and marbled effect. Most creamers show the cow alone, but some have a calf alongside, and the most desirable types have a seated milkmaid.

Pratt Lion
c.1800

At a Glance

Date: c.1800
Origin: England
Brief description: A Pratt type flatback model of a lion.

Felix Pratt (1780–1859) established a pottery at Fenton, Staffordshire and produced a wide range of useful and decorative wares. His name, however, will always be associated with a type of earthenware with distinctive orange, blue and green glazes and bright underglaze colors, applied by stippling or sponging, often over molded relief decoration.

This effect was widely imitated and therefore Prattware is the term applied to pottery of this type, whether made at Fenton or not. This Pratt type flatback model of a lion was probably made in Yorkshire where this style of decoration was particularly popular in the early 19th century.

BRITISH POTTERY AND PORCELAIN

Sir Isaac Newton Bust
c.1775

At a Glance

Date: c.1775
Origin: England
Brief description: Wedgwood Sir Isaac Newton bust modeled by Hoskins & Grant.

Josiah Wedgwood (1730-95) worked with several different potteries before setting up his own works at Burslem in 1759.

In his early years he produced black wares, agate, tortoiseshell and molded salt-glazed stoneware, before developing several highly distinctive forms which made his fortune. One of these was black basaltes, said to have been inspired by the basaltic columns of the Giant's Causeway. This fine stoneware was developed in the late 1760s and was extensively used for library busts of scientific, literary, artistic, and musical celebrities.

Much of the interest (and value) lies in the person portrayed, in this case Sir Isaac Newton. Hoskins & Grant modeled the bust from an original by François Roubiliac (c.1730) at the Royal Observatory, Greenwich, which was subsequently lost.

ANTIQUES AT A GLANCE: CERAMICS

Staffordshire Pearlware Group
c.1800

At a Glance

Date: c.1800
Origin: England
Brief description:
A Staffordshire pearlware bull-baiting group.

In 1779, Josiah Wedgwood devised a form of creamware, whose slightly bluish tinge suggesting a nacreous substance soon gave this pottery the name of pearlware. It was copied by other Staffordshire potters and continued to be popular until the middle of the 19th century, often serving as the basis for decorative hollow wares and figures.

This group shows the barbaric sport of bull-baiting, the tethered animal lunging at one of the hounds harassing him, while the "referee" stands behind with his hat upraised by way of encouragement. Such bull-baiting groups, dating around the beginning of the 19th century, are now much sought after.

BRITISH POTTERY AND PORCELAIN

Staffordshire Watch Holder Group
c.1870

At a Glance

Date: c.1870
Origin: England
Brief description: Two flat-back, hollow-molded, earthenware ornaments.

Staffordshire was renowned for its flat-back figures which attained the height of popularity in the late 19th century. These hollow-molded, earthenware ornaments with simple oval bases and distinctive flat backs evolved in the 1840s out of the flower vases and spill-holders that were made from the late 18th century onwards.

Kilted highlanders, wide-eyed milkmaids and naïve spaniels formed the bulk of these figures, but Royalty, national and international heroes and popular celebrities were increasingly portrayed. In addition to the figures intended as mantelpiece ornaments, the flat-back technique was applied to quasi-useful articles, such as these watch-holders dating around 1860–70 with figures in Highland costume reflecting a fashion inspired by Queen Victoria's frequent jaunts to Balmoral.

ANTIQUES AT A GLANCE: CERAMICS

Chelsea Jug
c.1745–49

At a Glance

Date: c.1745–49
Origin: London
Brief description: A Chelsea white "goat and bee" jug.

The earliest factories in England producing soft-paste porcelain were located at Bow and Chelsea in London and operated from the mid-1740s.

This Chelsea white "goat and bee" jug bears the incised triangle mark, which dates it to the earliest period (1744–49). The baluster form has a lower section molded with two recumbent goats below a loose spray of flowers applied with a bee, the branch handle applied with leaves. Chelsea wares of this period were often produced without colored decoration, the light, translucent body with its glossy texture is seen to its best advantage in this manner.

Chelsea Bocage Imperial Shepherds
c.1765–70

At a Glance

Date: c.1765–70
Origin: London
Brief description: A pair of Chelsea bocage figure groups known as the Imperial Shepherds, bearing the gold anchor mark of 1765–70.

In this period Chelsea was mainly inspired by Sèvres to produce a splendid range of decorative figures derived ultimately from the genre paintings of Antoine Watteau. The shepherdess and her rustic swain in these groups are modeled standing before a bocage of pink flowers, she with a lamb at her feet and he with his faithful hound at his side. Both figures are clad in elaborate brocade costume and the bases on which they stand are decorated in the finest Rococo fashion.

Bow Group
c.1752

At a Glance

Date: c.1752
Origin: London
Brief description: A Bow group of the fortune teller by the *Muses Modeler*.

A pottery was opened in the east end of London at Bow in 1744 by Thomas Frye, an Irish painter and by 1748 was producing soft-paste porcelain using bone ash, hence the name "bone china" which is sometimes applied to this form of porcelain. The body was more robust than that produced at Chelsea, and has a slight orange translucency when held up to the light. In the 1770s the firm was merged with that of Derby. The earliest wares imitated Chinese originals decorated in blue and white, but by the middle of the 1750s more distinctive styles of decoration and shape were beginning to emerge.

This Bow group shows a woman wearing a puce-lined, pale-yellow coat, white apron, and her dress painted and enriched in gilt with flower-sprays, a garland of flowers over her shoulder and with flowers in her hair. She holds out her hand to a bearded palmist wearing a pale-pink washed long-sleeved coat edged in puce and lined in pale-yellow, his iron-red boots enriched with gilding.

Chelsea Coffee Pot
c.1744–49

At a Glance

Date: c.1744–49
Origin: London
Brief description: A Chelsea coffeepot bearing the incised triangle mark of 1744–49.

The lobed sides of the baluster pot are crisply molded with a spiral flowering tea-plant. The high-domed cover has matching decoration. The handle has a foliage-scroll thumbpiece and the whole piece is mounted on an octafoil (eight-leaved) foot.

Derby Figures
c.1770

At a Glance

Date: c.1770
Origin: England
Brief description: A pair of Derby figures by Duesbury showing a boy and girl in rustic costume of the period, grooming a cat and dog respectively.

A factory producing soft-paste porcelain was opened at Derby about 1750 and continued until 1848—no connection with the Crown Derby factory which opened in 1876. After a period in which domestic wares predominated, Derby turned to more ornamental wares after William Duesbury took over as director in 1756, and began producing pot-pourri vases and salts in imitation of Meissen and Sèvres.

This pair of Derby figures by Duesbury date from about 1770 and show a boy and girl in rustic costume of the period, grooming a cat and dog respectively. Duesbury acquired the Chelsea works in that year, and subsequent products of this type are generally referred to as Chelsea-Derby.

Derby Teacup, Saucer and Sugar Bowl
c.1795

At a Glance

Date: c.1795
Origin: England
Brief description: A Derby yellow-ground teacup, saucer and sugar bowl painted by "Jockey" Hill.

The cup shows View at Little Chester Near Derby *and the saucer* Boat-house Near Derby, *within a gilt band and entwined ribbon cartouches. The bases have the crown, crossed batons and D marks in blue, indicating manufacture by William Duesbury, about 1795.*

Lowestoft Cream Jug, Coffee & Teapots
c.1775

At a Glance

Date: c.1775
Origin: England
Brief description: Lowestoft cream jug, teapot and coffee pot painted with Oriental figures in simple landscapes.

A pottery was established in the Suffolk seaside town of Lowestoft about 1757 and continued until 1802, producing soft-paste porcelain for local consumption. Although the bulk of its output consisted of domestic wares, it also produced some excellent decorative pieces, including some of the earliest souvenirs aimed at the nascent tourist market, artlessly inscribed "A Trifle from Lowestoft." Lowestoft was also caught up in the craze for chinoiserie in the 1770s and in this period produced teapots and coffee pots decorated with rather naïve but endearing Chinese subjects.

Worcester Deep Plate
c.1770

At a Glance

Date: c.1770
Origin: England
Brief description: Worcester deep plate with a large group of redcurrants and foliage at its center within a border of specimen fruit, including cherries, grapes and raspberries, divided by five blue and gilt C-scroll panels enclosing insects.

The longest-running of all the British porcelain factories, Worcester was founded in 1751 and put on a sound footing when it took over the well-established Bristol works a year later. Under the name of Royal Worcester it flourishes to this day. In the early years it concentrated on useful table and other domestic wares, but from 1760 onwards it diversified into more ornamental wares, at first painted in dark underglaze blue and then, from about 1770, decorated with a wider range of colors enhanced by gilding.

This deep plate dates from that decade and was produced as part of a service for the Duke of Gloucester.

ANTIQUES AT A GLANCE: CERAMICS

Worcester Serving Dish
c.1775

At a Glance

Date: c.1775
Origin: England
Brief description: This large rectangular serving dish also formed part of the famous Gloucester service.

The center is luxuriantly painted with peaches, an apple, cherries and grapes, while the border has four butterflies within blue and gilt cartouches divided by sprays of fruits.

BRITISH POTTERY AND PORCELAIN

Worcester Tableware
c.1770

At a Glance

Date: c. 1770
Origin: England
Brief description: Two Worcester blue-scale, cabbage-leaf-molded mask-jugs, two mugs, and a quatrefoil tureen.

The oviform body of the mask-jugs are reserved and painted with exotic birds perched on berried branches and with birds and insects within gilt scroll vase and mirror-shaped cartouches. The mugs are similarly pianted with exotic birds in landscape vignettes, but the tureen has swags of flowers.

Worcester Dessert Service
c.1820

At a Glance

Date: c.1820
Origin: England
Brief description: A Worcester part dessert service with molded gilt gadrooned rims and gilt seaweed borders, surrounding exquisite hand-painted vignettes of landscapes showing various notable churches and castles of England, Scotland, and Wales.

The longevity of the Worcester company has led collectors and ceramics scholars to divide it into a number of periods, according to the names of the various families that had a controlling interest. The period of the Flight family began in 1783 and combined with the Barr family a decade later, then Kerr and Binns (1852) and the present name adopted in 1862. This part dessert service belongs to the Flight, Barr and Barr period.

BRITISH POTTERY AND PORCELAIN

Pearlware Models
c.1780–95

At a Glance

Date: c.1780–95
Origin: England
Brief description: Pearlware models, including two parrots, two other birds, and a hen.

The figure of the bird (far left) is modeled in creamware, a fine-quality, cream-colored lead-glazed earthenware, which superseded delftware and rivaled European porcelain. The other figures are pearlware, similar in body to creamware but with a distinctive bluish glaze. All of these birds were produced in Yorkshire between 1780 and 1795.

Bristol Plaque
c.1775

At a Glance

Date: c.1775
Origin: England
Brief description: A Bristol biscuit oval plaque with a central medallion and initials.

The central medallion is inscribed in gilt with the initials SC within a burnished and matt gilt dentil cartouche, surrounded by richly gilt foliage. The initials are those of Sarah, sister of the manufacturer Richard Champion who produced this presentation piece about 1775.

Coalport Dessert Service
c.1805

At a Glance

Date: c.1805
Origin: England
Brief description: A part dessert service painted in the Imari style.

The Coalport porcelain factory on the banks of the Severn in Shropshire was established in 1795 and absorbed its neighbor and rival Caughley four years later. This part dessert service was painted in underglaze-blue, iron-red, green and gilt with flowering chrysanthemum alternating with dark-blue panels showing iron-red flowerheads among gilt foliage. It was produced by Coalport about 1805 and is a fine example of their bone china decorated in the Imari style.

Minton Teapot
1846

At a Glance

Date: 1846
Origin: England
Brief description: A Minton teapot designed by Felix Summerly (Henry Cole).

The simple white globular body and cylindrical neck are offset by such decorative features as the ram's head finial, lion spout and goat's head handle. Founded in Stoke-on-Trent by Thomas Minton in 1793, this innovative pottery encouraged artists to design its wares.

The career of Henry Cole (1808–82) took off when he designed this award-winning tea service in 1846 and in the following years he left Minton to found Felix Summerly's Art Manufactory, which continued until 1850. For his work in organizing the Great Exhibition of 1851 he received a knighthood. He helped to found the Victoria and Albert Museum but he should also be remembered for inventing the Christmas card (in 1843).

ART AND STUDIO POTTERY

Linthorpe Vase
19th century

At a Glance

Date: 19th century
Origin: England
Brief description: A Linthorpe vase designed by Dr Christopher Dresser with impressed marks.

The Linthorpe Pottery was established in 1879 by John Harrison when his Sun Brick Works near Middlesbrough was faced with closure due to falling demand. Harrison joined forces with Christopher Dresser to create this art pottery, which only enjoyed limited success and closed down in 1890.

Dresser hired Henry Tooth, an Isle of Wight artist, as pottery manager, although he had no previous experience of ceramics. He came to pottery with a fresh and original outlook and his flair for interesting colors and glazes is reflected in the startlingly eclectic designs of Linthorpe vases which drew on Chinese, Egyptian, Greek, Roman, Indian, Islamic, Peruvian and Celtic originals. This art pottery endured for only a decade but its products are now very much sought after.

Ault Vase
19th century

At a Glance

Date: 19th century
Origin: England
Brief description: A large vase designed by Christopher Dresser decorated in relief with Indians.

Tooth left Linthorpe in 1882 to go into partnership with William Ault, and in the same year Christopher Dresser severed his connection with Linthorpe, which went into decline from then onwards. Tooth and Ault formed the Bretby Pottery at Woodville, south Derbyshire but as a result of a disagreement Ault left in 1886 and the following year established his own art pottery at Swadlincote. Ault faience specialized in larger items, such as bowls, vases, jardinières and ornamental pedestals, but the extravagant glazes were strongly reminiscent of Linthorpe.

Between 1892 and 1896 Christopher Dresser designed a number of pieces for Ault, such as this large vase decorated in relief with Indians covered in yellow and brown running glazes.

Doulton Lambeth Bowl
19th century

At a Glance

Date: 19th century
Origin: London
Brief description: A Doulton Lambeth stoneware bowl by Frank Butler.

The Doulton Pottery was established at Lambeth, London in 1818. It is best known for its industrial pottery, plumbing materials, bathroom fittings, and ginger-beer bottles as its specialties, but in 1871 Doulton revived the art of decorative salt-glazed stoneware and was in the forefront of the Arts and Crafts Movement. The encouragement given to the Lambeth School of Art led to the development of studio pottery in England. In the closing years of the 19th century elaborate vases and bowl were decorated by Frank Butler, inspired by Islamic, Indian, and Chinese motifs.

Doulton Lambeth Vase
19th century

At a Glance

Date: 19th century
Origin: London
Brief description: A Doulton Lambeth stoneware vase by Hannah Barlow.

Among the artists employed by Doulton's Lambeth pottery was Hannah Barlow, who produced a great many pieces of art pottery in the period from 1883 to 1890, mainly "Carrara," a dense white stoneware decorated with colored patterns inspired by the famous Italian marble. Typical of the vases with painted and incised ornament by Hannah Barlow is this one showing a classical style.

Burmantofts Toad Spill Vase
19th century

At a Glance

Date: 19th century
Origin: England
Brief description:
A Burmantofts toad spill vase made from biscuit-colored earthenware covered with a fine felspathic glaze.

Burmantofts was a type of faience named after the Leeds, Yorkshire suburb of Burmantofts, where the pottery of Wilcock & Company was located. This firm originally produced industrial and architectural tiles and earthenware, but from 1880 onwards began manufacturing decorative wares consisting of a very hard biscuit-colored earthenware covered with a fine felspathic glaze. At first used for decorative tiles, it was extended to vases, bowls, and other hollow wares, often in fanciful shapes, such as this toad spill vase.

ANTIQUES AT A GLANCE: CERAMICS

Belleek Elephant
19th century

At a Glance

Date: 19th century
Origin: Northern Ireland
Brief description: A Belleek model of an elephant, and tiles and ornaments with flowers and leaves.

A porcelain factory was established at Belleek in Northern Ireland in 1857 and continues in production to this day. In the late 19th century it produced openwork baskets with floral incrustations and later on figures with a distinctive iridescent glaze. It also produced animal figures, such as this model of an elephant (top), as well as tiles and ornaments combining flowers and leaves with intricate undercutting (below).

ART AND STUDIO POTTERY

Martin Brothers Jug, Vase, and Birds
c.1898–1914

At a Glance

Date: c. 1898–1914
Origin: London
Brief description: Two grotesque birds, a jug, and a vase by the Martin brothers.

The four brothers had their own studios in Fulham and Southall (1873-1914) and produced hollow wares, but are best remembered for humanoid birds. The eldest brother, Robert Wallace, trained as a sculptor and was greatly influenced by the Gothic Revival, but through his connections with the Lambeth School of Art he was employed by Doulton as a modeler before setting up his own studio. As well as the grotesque bird figures, the Martin Brothers produced vases and jugs with grotesque ornament, such as the sea creatures on the jug shown here.

ANTIQUES AT A GLANCE: CERAMICS

De Morgan Vase
1888–1897

At a Glance

Date: 1888–1897
Origin: London
Brief description: A William De Morgan pottery twin-handled vase painted in ruby luster with a seated griffin among stylized foliage.

William De Morgan (1832–1917) was a ceramicist inspired by William Morris with whom he established a pottery at Merton Abbey near London, but later he broke away to found his own studio pottery in Fulham. At this studio (1872–1907) he also produced luster wares and highly intricate patterns influenced by the Italian Renaissance. Several of his designs were based on particular animals such as the griffin. Here the beast has lost its hind legs, its body is depicted as a reptile's tail.

ART AND STUDIO POTTERY

Burmantofts Faience Vases
c.1885

At a Glance

Date: c.1885
Origin: England
Brief description: Two Burmantofts faience vases, decorated by Lewis Kander.

Burmantofts produced a wide range of earthenware, but are best known for their Anglo-Persian style, inspired by De Morgan, and used to decorate very large hall vases, chargers, and plaques. All known pieces were decorated by Kander and bear his initials.

ANTIQUES AT A GLANCE: CERAMICS

Moorcroft Vases
late 19th/early 20th century

At a Glance

Date: Late 19th/early 20th century
Origin: England
Brief description: An urn-shaped vase, a two-handled vase, and six smaller vases in Macintyre or Claremont patterns.

William Moorcroft (1872–1946) was the leading artist-potter in Staffordshire at the turn of the century; his works immensely inspired by plant forms. In 1898, William Moorcroft developed the art pottery department at James Macintyre & Co, but he left in 1913 to found his own pottery at Corbridge, Staffordshire, backed by the Liberty family. Many of his early wares were made for Liberty and bear their retail mark. He was later noted for his luster, flambé and distinctive matt glazes.

ART AND STUDIO POTTERY

Poole Pottery Dish Vase and Spill-holder
1930s

At a Glance

Date: 1930s
Origin: England
Brief description:
Poole Pottery polychrome dish, vase and waisted spill-holder from the Delphis range.

An art pottery was established by John and Truda Adams, Harold Stabler and Owen Carter, the principal of the Dorset tile manufacturer Carter & Company, and out of this developed the Poole Pottery, which continues to this day. Both dishes and hollow wares by the Adamses were decorated by Stabler and are in keen demand.

ANTIQUES AT A GLANCE: CERAMICS

Rookwood Vases
late 19th century

At a Glance

Date: Late 19th century
Origin: USA
Brief description: Three Rookwood earthenware vases with exotic glazes.

The foremost American art pottery of the late 19th century was Rookwood, which flourished in Cincinnati from 1880 until its closure in 1967. It was founded by the wealthy socialite Maria Longworth Nichols in a converted schoolhouse on her father's estate. Beginning modestly, it expanded its production of art pottery in 1883 and was the leading exponent of studio pottery in America until the end of the century. It produced earthenware with exotic glazes, often with molded bird or animal forms (such as the falcon with outstretched wings in the vase with a matte maroon, blue and turquoise glaze (center).

ART AND STUDIO POTTERY

Roseville Twin-handled Bowls
c.1900s

At a Glance

Date: c.1900s
Origin: USA
Brief description: Roseville twin-handled bowls with applied relief molded ornament.

A pottery was established at Roseville, Ohio in 1892 and produced useful domestic wares until 1900 when it expanded into the realm of art pottery under the direction of George F. Young, emulating the products of Rookwood. The Rookwood art pottery was distinguished by the name "Rozane," usually incised on the base of each piece. Artistically, Roseville was not as accomplished as its great rival, but this was balanced by the originality and diversity of its styles, notably the twin-handled bowls with very unusual shapes and applied molded relief ornament.

ANTIQUES AT A GLANCE: CERAMICS

Weller Dickensware Vase
early 20th century

At a Glance

Date: Early 20th century
Origin: USA
Brief description: Weller Dickensware vase decorated with a panel of a female golfer.

Samuel A. Weller (1851–1925) founded a commercial pottery at Zanesville, Ohio in 1882 but gradually diversified into large decorative wares. His attention was drawn to studio pottery at the Columbian Exposition in 1893, and subsequently he took over the Lonhuda Pottery in Steubenville. Louwelsa pottery (a name compounded from the name of his daughter Louise and his own initials) continued until 1949 when the factory closed.

Other decorative wares were marketed under the name of Sicardo (after the French artist-potter Jacques Sicard) and Dickensware, inspired by the writings of the English novelist. A typical example of the latter is this slender vase with a motif of a Victorian lady golfer.

Grueby Pottery Vase
c.1899

At a Glance

Date: c.1899
Origin: USA
Brief description: An earthenware vase by the Grueby Pottery, incised with the decorator's monogram GB.

Grueby faience was a distinctive brand of American art pottery produced by a company that flourished briefly at East Boston between 1897 and 1910. The founder, William H. Grueby, had previously manufactured tiles and commercial wares but was inspired to diversify into art pottery after seeing the matte glaze pieces by the French ceramicist Auguste Delaherche at the Columbian Exposition of 1893. Grueby himself perfected an opaque enamel glaze, which attracted enormous attention on both sides of the Atlantic. Grueby made many of the bases used for Tiffany lamps.

Amphora Bowls, Vases and Jugs
c.1900

At a Glance

Date: c.1900
Origin: USA
Brief description:
A group of earthenware bowls, vases and jugs by Amphora.

The body of these wares is decorated in iridescent green, violet and pink glazes with gilt-enameled detail and curvilinear handles and ornament in the finest tradition of Art Nouveau with the distinctive Amphora Elite seal on the bases.

This American company took its name from the famous two-handled wine-jugs of ancient Greece and much of their wares were neo-classical in style.

ART AND STUDIO POTTERY

Bernard Leach Box and St Ives Chargers
early 20th century

At a Glance

Date: Early 20th century
Origin: England
Brief description: A blue and white porcelain circular box and cover by Bernard Leach.

Bernard Leach (1887–1979) and Shoji Hamada (1894–1978) worked in Japan before settling at St Ives, Cornwall where they established a studio pottery producing stoneware, earthenware and porcelain.

ANTIQUES AT A GLANCE: CERAMICS

Shoji Hamado Dish
early 20th century

At a Glance

Date: Early 20th century
Origin: Japan
Brief description: A caramel-glazed large stoneware dish with a blue cross motif by Shoji Hamado.

The dish comes in a wooden presentation case signed on the inside of the lid Shoji with the chop (seal) Sho and inscribed in Japanese "Ame-yu ao juyi gake moribachi," which translates as "bowl with caramel glaze and blue cross design."

ART AND STUDIO POTTERY

Bernard Leach Charger
c.1940

At a Glance

Date: c.1940
Origin: England
Brief description: An early stoneware charger by Bernard Leach, decorated with slip-trailed tree and animal motif with impressed BL and St Ives mark.

Under the influence of Shoji Hamada, Bernard Leach strove to produce functional forms simply decorated in motifs inspired by the arts of Japan, China and Korea.

St Ives Plate
1940s

At a Glance

Date: 1940s
Origin: England
Brief description:
A St Ives stoneware plate, with the impressed mark of that studio pottery.

Although the plate lacks the BL monogram it has been attributed to Bernard Leach because it is very characteristic of his style, that of a very simple dish covered in a mushroom glaze with brushwork decoration.

Hans Coper Dish and Vase
c.1946–58

At a Glance

Date: c.1946–58
Origin: England
Brief description: A stoneware bowl and "Poppy" vase by Hans Coper (1920–81).

From 1946 to 1958 Hans Coper shared a studio with Lucie Rie (1902–95) who had been trained in Vienna. They specialized in stoneware with complex glazes and textures.

Hans Coper Vase
c.1970

At a Glance

Date: c.1970
Origin: England
Brief description: A black-glazed stoneware vase by Hans Coper.

Coper's pottery is sculptural, and relies on texture and form rather than the conventional use of colored ornament. This vase is of diabolo form on a cylindrical foot, with the artist's HC monogram in an oval on the base.

ART AND STUDIO POTTERY

Clarice Cliff Dinner Service
c.1930

At a Glance

Date: c.1930
Origin: England
Brief description: A part dinner service designed by John Armstrong and decorated by Clarice Cliff.

This part dinner service designed by John Armstrong includes two tureens and covers, three large oval plates (one shown), dinner, dessert and side plates (one of each shown), decorated by Clarice Cliff in one of her "Bizarre" patterns, about 1930. Crude but colorful, they epitomized the Jazz Age.

Index

Adams, John and Truda 131
Aesthetic Movement 16
Amphora Elite 136
Andries, Jasper 20
Art Deco 11, 17
Art Nouveau 11, 16, 17
Arts and Crafts Movement 11, 16, 25, 123
Ault, William 122
Barlow, Florence and Hannah 25, 124

Baroque 13, 14
Bauhaus 17
Belleek 24, 126
Berlin 22
Biedermeier 16
Bing & Grondahl 22
blanc de Chine 21
Böttger, Friedrich 22, 76
Bow 14, 24, 108
Breslau 76
Bretby 122
Briggle, Artus van 25
Brislington 20
Bristol 13, 20, 118
Burmantofts 125, 129
Burne-Jones, Edward 16

Cardew, Michael 25
Carter, Owen 131
Castelli 31, 33
Caughley 14, 24
Chelsea 8, 14, 22–4, 106-7, 109
Chinese porcelain 21–2, 50–69
Cliff, Clarice 25, 143
Cloisonné wares 48–9

Coalport 14, 24, 119
Cole, Henry 25, 120
Cooper, Susie 25
Copeland 24
Copenhagen 22
Coper, Hans 25, 141–2

Davenport 24
de Morgan, William 25, 128
Deck, Theodore 25
Dedham 25
Delamain, Henry 41–2
Delft 13, 20–1, 35–40
Delftfield 13
Della Robbia 25
Derby 14, 20, 24, 110–11
Deruta 26, 29
Dickensware 134
Dihl and Guerhard 93
Doccia 86
Doulton 24, 25, 123–4
Dresden 22
Dresser, Christopher 121–2
Du Paquier 87
Dublin 13, 20, 41–2
Duesbury, William 24, 110

Faenza 13, 20, 30
Faience 20, 28–32, 35, 44
Familia Gotica 30
famille jeune 21
famille noire 21
famille rose 21, 50–4
famille verte 21, 55, 59–63
Frankenthal 22
Frankfurt 32
Fritsche, Elizabeth 25

Fry, Laura 25
Frye, Thomas 24, 108
Gardner 22, 94–5
Georgian 13, 14
Glasgow 20
Gothic style 14, 16
Grondahl, Bing and 22
Grueby Faience 25, 135

Hamada, Shoji 25, 137–8
Harrison, John 121
Heath, John 24
Heylyn, Edward 24

Imari 46–7
Istoriato 27–8

Jiajing period 57
Johnson, Jacob 20
Jugendstil 11, 16

Kaendler 22
Kakiemon 45
Kangxi period 58–63, 69
Kornilov 22
Kuznetsov 22

Leach, Bernard 25, 137, 139-40
Leeds 14
Liberty style 11
Lille 44
Linthrope 121
Limerick 13, 20
Liverpool 13, 20
London 13, 20
Longton Hall 14
Lonhuda 25, 134
Louwelsa 25
Lowestoft 14, 112

Ludwigsburg 22
Luxembourg 97
Maiolica 13, 20, 26, 28, 29, 30, 34
Majolica 13
Martin Brothers 127
Mason's ironstone 24
Massier, Clément 98
Meissen 10–11, 14, 20, 22, 70–82
Mettlach 96
Ming period 64–6
Minton 14, 24–5, 120
Moorcroft 25, 130
Morris, William 11, 16
Moscow 22

Nantgarw 24
Naples 84-5
Naturalism 17
Neo-classical style 14
New Hall 14
Nuremberg 43
Nymphenburg 22, 43, 83

Qianlong period 54
Quinze, Louis 20

Paris 93
Pearlware 104, 117
Persian wares 7
Pesaro 28, 34
Poole 24, 131
Popov 22, 94
Pratt, Felix 102

Rathbone, Harold 25
Rie, Lucie 25
Rockingham 14, 24
Rococo 14, 21

Rookwood 25, 132
Roseville 25, 133
St Ives 137, 139–40
St Petersburg 22, 94
Second Empire 16
Sèvres 14, 22, 88–92
Sezession movement 17
Spode 14, 24
Sprimmons, Nicholas 24
Stabler, Harold 131
Staffordshire 24, 99–101, 104-5
Studio pottery 24–5, 137–42
Swansea 14, 24
Symbolism 17

Urbino 13, 20, 27, 29

Vienna 22, 87
Villeroy & Boch 97
Vincennes 22, 88

Wall, John 24
Webb, Philip 16
Wedgwood 14, 24–5, 103
Weller, Samuel A 25, 134
Wemyss 24
Wincanton 13, 20
Worcester 14, 20, 24, 113–16

Yaoling Zun 58
Yuan period 56
Yongzheng period 52
Yorkshire 125